W9-CNR-073

Bioethics in America

Bioethics in America

ORIGINS AND CULTURAL POLITICS

M. L. Tina Stevens

The Johns Hopkins University Press
Baltimore and London

©2000 M. L. Tina Stevens
All rights reserved. Published 2000
Printed in the United States of America on acid-free paper
9 8 7 6 5 4 3 2 1

The Johns Hopkins University Press
2715 North Charles Street
Baltimore, Maryland 21218-4363
www.press.jhu.edu

A catalog record for this book is available from the British Library.

Library of Congress Cataloging-in-Publication Data
Stevens, M. L. Tina.
    Bioethics in America : origins and cultural politics / M. L. Tina
Stevens.
        p.   cm.
    Includes bibliographical references and index.
    ISBN 0-8018-6425-9 (hardcover : alk. paper)
    1. Medical ethics—United States—History.   2. Bioethics—
United States—History.   I. Title.
R724.S84 2000
174′2′0973—dc21   00-008389

To my family,

especially to my husband,
Stephen Shmanske,

and to the memory of my parents,
Richard Stevens
and
Gloria Keesey Stevens

Contents

Preface

Like many physicians, Dr. McIntyre wore an electronic beeper that alerted him when he was needed for medical consultation. But Dr. McIntyre was not a physician, nor did he give medical advice. He was a Lutheran minister and philosopher who provided ethical counsel for five New Jersey hospitals. Should this defective newborn undergo surgery? Should that respirator be discontinued? It was 1979 and Dr. McIntyre was a bioethicist, on call to advise on questions such as these.[1]

During the 1960s, a diffuse, self-selected group of philosophers, theologians, lawyers, scientists, and doctors began to examine the ethical implications of biomedical technologies that, for them, held the possibility of transforming the moral order. Eventually earning the appellation *bioethicists*, these intellectuals and practitioners maintained that society was unprepared to answer the moral questions posed by novel technologies such as in vitro fertilization, amniocentesis, organ transplantation, the artificial respirator, genetic engineering, and genetic screening, and that, therefore, their services were needed.

It became typical for a hospital to employ an interdisciplinary "ethics committee." In 1972 the ethics review board at the Johns Hopkins Hospital, for example, comprised a pediatrician, a surgeon, a psychiatrist, a clergyman, and a lawyer. Oftentimes a philosopher was part of the mix. Even before the Johns Hopkins ethics cabinet began its work, it had gained the sobriquet, "God Committee."[2] In 1987 the American Hospital Association estimated that there were ethics consultants or committees at 60 percent of acute care institutions and at 80 to 90 percent of major medical centers.[3]

Bioethics institutes, committees, commissions, and courses continue to flourish.[4] In 1992 bioethicists celebrated thirty years of bioethics with a special Birth of Bioethics conference in Seattle, Washington. With a burgeoning publishing activity worth millions and firm institutional moorings in universities, law schools, medical schools, and hospitals, and in media coverage, bioethics is not simply a discernible feature of recent biomedical practice. It has become a conspicuous American cultural fixation.

This book explores the historic cultural roots and social ramifications of bioethics as a postwar phenomenon. With its most prominent institutions emerging during the late 1960s, bioethics is viewed mainly as a product of that turbulent decade, either as a response to uniquely problematic technologies or as one of many defiant challenges to authority and central social institutions. But bioethics was not simply the spontaneous creation of the 1960s. It was, instead, a recent expression of a centuries-long cultural legacy of American ambivalence toward progress. The Prologue offers a brief explanation of this phenomenon. Throughout U.S. history, such ambivalence has served as midwife to technologies and processes of technological development, which, on first impression, may disconcert some observers and even some consumers. This ambivalence—following a path of generational emergence and subsequent dissipation or absorption—has been one of the pillars of a stable American technological society. In this sense, when bioethics seemed not to dissipate but instead secured institutional anchors at the close of the 1960s, it appeared to be a historical aberration. In fact, bioethics had merely become institutionally congealed: an outward sign of an elite anxiety that won institutional legitimacy because it had proved far less threatening to existing social arrangements than the changes demanded by more radical, and more popular, social critics of the sixties.

Chapter 1 presents the specific historical circumstances that gave rise to the postwar incarnation of this ambivalence, that is, bioethics. I evaluate the fate of the tradition of ambivalence in the wake of post-atomic cultural politics and conclude that, just as the cultural taproot of bioethics is not found in the 1960s, neither are its more recent sources. The contemporary roots of bioethics stretch back, instead, to the postwar "responsible science movement" of the 1950s. Contrary to prevailing wisdom, bioethics did not begin as a nonprofessional call for critique and control of science and medicine. It began when intellectuals heeded the plea of biomedical researchers for interdisciplinary and "lay" scrutiny of their work. Chief among these researchers were the postwar geneticists. The aim of some of these genetic researchers was to avoid the social and moral disquietude suffered by the atomic scientists of the 1950s; the aim of others of them was to stave off the possibility of more virulent external control of various eugenic proclivities and proposals. The glare of social commentary from the late 1960s and the 1970s, which pitted scientific and medical interests squarely against "the people," has made it difficult to recognize the significance of appeals from within science and medicine calling for the dissemination of erudite information and seeking external scrutiny (but not control) of professional research and behavior.[5]

I also consider the influence of more radical social critics. During the 1960s, the public evaluated calls from the responsible science movement against a background of analyses that were more threatening to existing social arrangements. Critics like Lewis Mumford, Jacques Ellul, Herbert Marcuse, and Theodore Roszak perceived far-reaching interconnections between science and other social institutions. They called for more than simple caution and oversight; they desired fundamental change of a politically radical nature. How these more radical critiques affected bioethical development and helped to establish bioethics' cultural authority are addressed in this chapter.

Finally, Chapter 1 focuses on bioethical literature itself, uncovering its modern roots in the post–World War II responsible science movement and considering it within the broader context of American intellectual culture in the 1960s and

1970s.[6] In so doing, the chapter reveals that dominant views of the history and social function of bioethics misidentify both the temporal genesis and historic social function of bioethics. Locating bioethics' latter-day roots more accurately in the 1950s, this chapter suggests that while bioethics was not simply another antiauthoritarian impulse that surfaced in the 1960s, it was, nevertheless, a complex participant in the cultural politics of that chaotic decade. It was a phenomenon that confronted the more self-interested institutional values of medicine and science while simultaneously recreating the legitimacy that ensured the longevity of those values. Bioethical impulses found their way into enduring social institutions not because they represented the social challenges of the 1960s but because they successfully diffused those challenges.

In Chapter 2 I explore the impulse within the bioethics movement to professionalize and assess the vitality of this impulse as it strained against the opposing inclination to remain a diverse lay social force. The history of the Hastings Center in New York, the nation's first bioethics institute, is a poignant story of the challenges facing pioneering bioethicists in their struggle to achieve institutional longevity while safeguarding autonomy. Investigating the Hastings Center means evaluating the success of the bioethics movement in maintaining an "ethical independence" from biomedical political interests. The story is an example of how funding institutions penalized early bioethicists for assuming a posture too critical of biomedicine. In this manner, independent oversight was compromised. As temperate views were rewarded (thereby facilitating institutionalization), radical critiques were diffused.

Some of the consequences of a compromised oversight and radical diffusion are explored in Chapter 3's analysis of the redefinition of death—one of the first issues to be addressed systematically by early bioethicists. I discuss how, contrary to the declaration of the promulgators of the brain death definition of death, the mechanical ventilator did not necessitate redefining death. Instead, the impulse to redefine death surfaced when the demands of researchers for transplantable organs created a conflict of interest on the part of doctors when deciding whether to "pull the plug" on potential organ donors who were irreversibly comatose. I examine how transplantation researchers relied on bioethicists to assist them in their efforts to manage public opinion.

The unanticipated consequence of the redefinition of death led to the 1975 Quinlan "death with dignity" case—a case which, as Chapter 4 argues, created an unprecedented demand for bioethical opinion and significantly contributed to establishing the cultural currency of bioethics. When Karen Quinlan slipped into a coma in the summer of 1974, the ensuing controversy surrounding this "sleeping beauty" riveted the American public and increased the demand for bioethical scrutiny of professional behavior. Many Americans saw in Karen's misfortune a frightening specter of a medical profession which did not ease suffering but which, instead, extended agony. I describe popular misconceptions surrounding the controversy—misconceptions that fueled the growth of bioethics. I find that the Quinlan case, often seen as the quintessential example of how external rules curbed professional discretion, was actually less a gift of freedom from "rampant technology" to patients like Karen Quinlan than a gift of freedom from liability to organized medicine.

The bioethics "movement," I conclude, assisted in transforming alarm over exotic technologies into a situation in which ethical experts manage problems—problems generated by technologies seen, ironically, as value-neutral in their creation even while they are problem-causing in their outcomes. By the mid-1970s, bioethics was functioning as midwife to technologies and to a medical research community in need of broad social acceptance. The history of bioethics, considered within the larger narrative of 1960s protest culture, suggests that the movement from the sixties to the seventies was a shift from critique to management. The cultural politics of bioethics exemplifies an important phenomenon in recent American history, namely the ebbing of the protest culture of the 1960s.

Acknowledgments

Special thanks to Daniel Callahan for providing access to Hastings Center archives and to both him and Willard Gaylin for allowing me to take their oral histories. I am indebted also to the following bioethicists who allowed me to interview them: Thomas Beauchamp, Bruce Jennings, James Nelson, and Robert Veatch. My thanks as well to the Rockefeller Archives Center, especially Thomas Rosebaum, for his kind assistance.

I offer thanks to University of California–Berkeley's Paula Fass and Thomas Laqueur for crucial validation of nascent ideas and insistent challenges; and especially to Carolyn Merchant, from U.C. Berkeley, for kindness, rigor, and friendship. I am grateful, too, to U.C. Berkeley's Jim Kettner, Sam Haber, Jack Lesch, and David Hollinger for either encouragement or important comments on manuscript fragments, and to Jaqueline Wehmueller for being such a wonderful editor. Although it is no longer possible to thank Jack Pressman, formerly of University of California–San Francisco, I am grateful to be able to acknowledge his collegial friendship, humor, and special insight, which continue to be missed. Early on, Shelly Messinger, Marge Schultz, and Martin Shapiro from U.C. Berkeley's Jurisprudence and Social Policy Program gave helpful comments or general encouragement for which I extend my thanks.

Continuing gratitude and affection go to my friends and colleagues who lent either crucial support or valued judgment: Jesse Barrett, Karen Bradley, Byde Clawson, Pat Conolly, Beth Haiken, Julia Rechter, Susan Spath, Jessica Weiss, and Lisa Zemelman. Additional thanks to Rosann Greenspan, for demystifying intellectual puzzles; to Patti Martin, for helping hatch an interest in bioethics and seeing it through to the epilogue; and to Celeste Calvo, for expert transcriptions of oral histories.

With love and pleasure I thank the members of my exceptional family for treasured affection and support (both the earnest and teasing varieties): Richard O. Stevens, Esq., Gloria Jeanne Stevens, William A. Stevens, Joan Ryan Stevens, Sylvia and Bob Gadient, Gerry Keesey Hoppe, Elizabeth and Bernard Shmanske, and the Stevens urchins: Elise, Christina, Ryan, and Paul. Even after the deaths of my cherished parents, Richard Stevens and Gloria Keesey Stevens, their pride in me sustains my work. For this, and for their love which has been undying, I am grateful.

Going beyond familial affection, two siblings extended important professional judgment: Peggy Stevens contributed significant editorial skill; James E. Stevens, Esq., offered insightful comments and ensured, with infectious interest, that I attend to relevant articles.

Final thanks are for my husband, Stephen Shmanske, for the impressive critical intelligence that he brought to bear on my work, and for his frequent encouragement, unwavering devotion, and abiding love.

Bioethics in America

The Tradition
of Ambivalence

Although much about bioethics was a unique response to the social and cultural conditions of its time, an essential feature of the original bioethical profile—its guardedness against unfettered biotechnological production—is descended from a lengthy tradition of ambivalence toward "progress." This cautionary outlook forms a substratum in American thought; it is a circumspection lying just beneath an otherwise overtly celebratory national cast of mind. The specific intellectual and cultural milieu giving rise to this hesitancy has varied from generation to generation, but its social location has remained constant: it has emanated from among the educated elite whose larger social class fundamentally supports scientific, medical, and technological development.

The degree of hesitancy exhibited by early bioethicists and their predecessors ranged from caution to ambivalence. Often these critics were anxious over the more dispiriting implications of technological "advance," and sometimes they were wary of losing control over a seemingly autonomous process. Typically, they did not indict producers or controllers of technology. Some of them were careful to discuss the importance of class in their work in other areas. With respect to problems associated with technology, however, rarely did they identify a responsible class, a "system," or an establishment of technological creators or an elite coterie with vested interests in research and development. They were more apt to discuss "man's" confrontation with an ineluctable process—a process politically disembodied from its productive sources—and what "our" response (i.e., our culturally homogenous and socially classless response) should be as individuals.

The recurring emergence and dissipation of this ambivalence suggests that this cycle is an essential feature of a successful technological society: the process of dissolution signals the acceptance of technologies that, when first introduced, are unsettling. In this sense, when bioethical concerns seemed not to dissipate but instead were institutionalized at the close of the 1960s, bioethics appeared to be historically exceptional. Bioethics won this legitimacy, however, as a reward for functioning as a kind of cultural inoculation, an immunization that forestalled the more virulent attacks of radical critics, who were mistrustful of biomedicine's undergirding role in a technological society.[1] Contrary to dominant contemporary accounts of the history of bioethics, the field did not develop out of hallmark 1960s impulses to wrest control from medical and scientific institutions or simply as a unique response to exotic technologies; bioethics developed from critical sources emerging from within medicine and science and from the tradition of ambivalence more generally. The tradition of ambivalence provides a context for considering the place of bioethics within this social and historical phenomenon.

Given America's heady optimism for growth and development, it is easy to forget that there have been many who, while not against technological development per se, were nevertheless hesitant about the cultural costs associated with the technological advance. Rather than a stark division between an orthodox posture toward technology and a radical critique of this orthodoxy, there has been a spectrum of attitudes toward technology with unique political consequences; liberal (not radical) critics served to buttress the political infrastructure of technological development despite their criticism of it.

American enthusiasm for the forward-looking aspect of prosperity often has drowned out the voices of ambivalence, voices in evidence since the days of early settlement. Though not directed specifically at technology, for example, Puritan guilt established a firm undergirding of a general caution toward material success and of vigilance toward the "evils" that can accompany it.[2]

In postrevolutionary America, those discomfited by the uneasy alliance between republican ideology and industrialization directed their concern more specifically.[3] Thomas Jefferson's early hopes for the young republic, for example, left little room for market-industrial development. He believed with others of his age that, throughout history, republics had met their demise through the ills associated with the excesses of material development: Jefferson wished to discourage domestic manufactures in the new American republic and leave the menaces associated with industrialization to Europe. "Let our work-shops remain in Europe," Jefferson urged, "the mobs of great cities add just so much to the support of pure government, as sores do to the strength of the human body. It is the manners and spirit of a people which preserve a republic in vigour."[4]

Despite this inclination, Jefferson and other like-minded republicans were unable to resist the technological lure of increasing prosperity. Eventually, industrial promoters found ways of melding republican ideology with the idea of advancing technology—by arguing, first, that the republic could not truly be secure unless it were independent of Europe and, ultimately, that the luxury produced by prosperity was not an evil but rather, as Ben Franklin claimed, a "great spur to labour and industry."[5]

Ultimately, Jefferson's pastoral vision (a vision he himself did not realize) could not withstand the force of change. By the nineteenth century, the transportation and market revolutions had become a quickening reality.[6] As the nation sped its way through these transformations, however, Puritan and republican wariness did not vanish but assumed a position of dissent, a position evidenced most distinctly in the writings of Henry David Thoreau.

Thoreau evinces the ambivalence that some exponents of American transcendentalism felt toward the "progress" of material development. "What is the nature of the luxury which

enervates and destroys nations?" asked Thoreau in 1854. Almost as a response to Jefferson's desire that manufactures remain in England, Thoreau declared his disbelief that the "fac-
tory system is the best mode by which men may get clothing.
The condition of the operatives," he suggested, "is becoming
every day more like that of the English." But Thoreau also revealed a reluctant attraction to the urban and industrial development of "civilized" society. Not intending that humankind
should abandon the benefits of progress altogether, Thoreau
advised that "it certainly is better to accept the advantages,
though so dearly bought, which the invention and industry of
mankind offer." "I am refreshed and expanded," he mused,
"when the freight train rattles past me." Even so, the "civilized" path to industrial and urban development, Thoreau believed, had led the mass of men to lives of "quiet desperation."
Despite the dispiriting effects of development for Thoreau and
some of his fellow transcendentalists, however, urban and
technological advance were firmly a part of dominant American cultural life by the 1850s.[7]

By the late nineteenth century's "second industrial revolution," Americans' spirited belief in the virtues of industrial
technology was a cultural feature of unprecedented scope. The
century after 1870 was, according to one historian, "the American genesis," when Americans created a modern technological nation, characterized by mass production, the burgeoning
of industrial cities, and the growth of corporations. It included
as well the development of higher education devoted to technical education and a class of experts—engineers and industrial
scientists. "A nation of machine makers and system builders,
[Americans] became imbued with a drive for order, system, and
control."[8]

Yet, even from within this dominant celebratory milieu,
many educated Americans were uncomfortable with the has-
tening pace of growing industrialization. During this era of
modernization, the persistent anxiety that accompanied development was not so much stilled as transformed. Between 1880

and 1920 (a period, according to T. J. Jackson Lears, marked by "a national chorus of self-congratulation"), a diffuse group of individuals from the upper reaches of society—academics, journalists, literati, and ministers—revealed its uneasiness with the dominant culture. In many respects, these "antimodernists" were forerunners, culturally and intellectually, of bioethicists. Whereas bioethicists were to give direct and conscious expression to their disquietude, however, antimodernists often turned to alternative cultural expressions like medievalism, orientalism, and "primitivism"—a symptom of their discomfiture with the modern technological nature of society. Antimodernist interests expressed an uneasiness that was pervasive throughout the middle and upper classes; and antimodernist sentiment was ambivalent: critical of material development but also, at times, eager for it. Lears explained how this "half-commitment" assisted modern culture in defusing antimodernist dissent:

> Half-committed to modernization, antimodernists unwittingly allowed modern culture to absorb and defuse their dissent. Unable to transcend bourgeois values, they often ended by revitalizing them. Ambivalent critique became agenda for bourgeois self-reformation: antimodern craft ideologues became advocates of basement workbench regeneration for tired corporate executives; antimodern militarists became apologists for modern imperialism ... Antimodern thinkers played a key (albeit often unknowing) role in revitalizing the cultural hegemony of their class during a protracted period of crisis.[9]

Thus, although disturbed by materialism and technological and industrial development, antimodernists, through their ultimate absorption into modernist culture, furthered that milieu and, in so doing, eased the passage to a more secular and corporate world.

Aided by the dissipation of antimodernist angst and vitalized by the antimodernists' transvaluation of older, inhibiting bourgeois values, an unparalleled flood of consumer goods inundated American homes during the early decades of the twentieth century.[10] Promoters touted the labor-saving and pleasure-enhancing benefits of appliances, radios, and automobiles. By the interwar years, America had reached the zenith of its "technological enthusiasm," an era in which the United States was the acknowledged leader of the modern technological world.[11] The motto of the 1933 Chicago Century of Progress Fair characterized succinctly the dominant cultural milieu: "Science Finds—Industry Applies—Man Conforms."[12]

Intellectuals attempted to estimate the cultural effects of this stunning technological progress.[13] While the essayists in Charles Beard's edited volume *Whither Mankind: A Panorama of Modern Civilization* directed attention to a variety of areas, the "problem" of technology figured preeminently in their concerns. The West was a machine culture and American civilization was, Beard explained, "the full flower of the machine apotheosized." And, far from dissipating, "technological civilization ... threaten[ed] to overcome and transform the whole globe."[14]

Beard, although cautious, was not a naysayer to technology or "the machine." While it was true, he estimated, that "the machine system ... present[ed] shocking evils and indeed terrible menaces," he felt nevertheless that, "under the machine and science, the love of beauty, the sense of mystery, and the motive of compassion ... are not destroyed." Beard and his contributors did not call for a halt to any technological advance or for a rearrangement of machine civilization per se. Rather, he petitioned simply for study of society's technological processes so that "mankind" could better control them: "by understanding more clearly the processes of science and the machine," he concluded, "mankind may subject the scattered and perplexing things of this world to a more ordered dominion of the spirit."[15]

Economist Stuart Chase also steered a judicious, if more critical, course in considering the social ramifications of technological change. In his popular 1929 book *Men and Machines*, Chase discussed the pros and cons of technology, taking in turn the philosophies of "ecstasy, of gloom and of the fence rail." In the final analysis, Chase revealed his inclination "to the belief that machinery has so far brought more misery than happiness into the world."[16] Change always held open the possibility of improvement, however, and perhaps the triple menaces of "war, technological tenuousness and failure of natural resources" could be forestalled. But humankind had to "face the full implications of the machine." In language foreshadowing the bioethical ethos three decades in the future, Chase urged that while "man is not the slave of his machines . . . he has allowed them to run unbridled, and his next great task is, by one method or another, to break them to his service."[17]

The "responsible science movement" that followed World War II transformed the tradition of ambivalence once again. When fears surrounding technological development (in this case, the atomic bomb) were galvanized most apparently by scientists themselves, a renewed urgency gripped the reading public. Not merely ambivalent, postwar social critics of a more radical bent emerged, critics who were horrified by the stealthy production of the atomic bomb and the ensuing mass destruction. Their horror compelled them to identify political sources of technological production and to suggest political responses for control of technological development. Bioethics grew between the narrow strait of this new, more radical hostility toward science and the legacy of traditional ambivalence. How bioethicists, the cultural descendants of ambivalent social commentators from Jefferson through Thoreau to Chase, rose into bold relief against this recast background is taken up in Chapter 1.

Chapter One

The Culture
of Post-atomic
Ambivalence

In 1977, Hastings Center founder Daniel Callahan characterized bioethicists as trying to wrench control from physicians.
"Doctors want . . . to make all the choices," he said. "Well,
we're saying to them, no. There are principles you have to abide
by. . . . You're playing in a public ball-park now . . . and you've
got to live by certain standards . . . like it or not."[1] But almost
fifteen years later, he reflected soberly on the role that physicians themselves had played in forming the "dialogue" of
bioethics. "We were able to get doctors to help us form the
Hastings Center," he explained, "because they realized that
the dialogue had to encompass more than doctors. . . . A number of physicians were beginning to say among themselves,
'Look, we are getting more and more of these problems that
are harder to deal with.'" Callahan nuanced the formative
years in bioethics as constituting an alliance between a small
number of "whistle-blowing" doctors and "outsiders."

> I think the role of [anesthesiologist] Henry Beecher was very im-
> portant in the mid 60s when he blew the whistle on some bad ex-
> periments with human subjects which did real harm. . . . Beecher
> in fact said, "Look, we can't trust ourselves any longer." Some of
> these researchers do bad things. And Congress came along and es-
> tablished the Institutional Review Board Committee because the
> scientists said, "We don't need review, we are wonderful, we never
> did harm anybody" and in fact the government said, "We don't
> believe you anymore. You've got to have some public oversight."
> . . . I suppose the net effect of all this was that (a) you've got a
> group of people, doctors and others, that are interested in this

field; and (b) doctors couldn't keep these issues to themselves much any longer partly because some doctors were blowing the whistle saying, "Hey public, watch, look here, pay attention. There are some things going on here that are not good and that you should know about." This was probably a very small minority group of doctors. But the doctors were long resistant to the idea of any outsiders coming in.[2]

Callahan's recollections recognized that at least some part of the impetus for the development of bioethical guardedness toward biomedical advance came from within medicine during the 1960s and 1970s. Taking a good look at whose work early bioethicists were reading and how they came to be concerned with what some were calling the "biological revolution" offers a clear picture of when and from where bioethical impulses first arose. Such an examination places the emergence earlier than the sixties. It also brings into bold relief how the aims, celebrity, and influence of the atomic scientists were significant elements directing the development of bioethics.

"Probably terrible wars will arise and prove costly in blood and treasure," surmised Charles Beard in 1928, "but it is a strain upon the speculative faculties to conceive of any conflict that could destroy the population and mechanical equipment of the Western world so extensively that human vitality and science could not restore economic prosperity and even improve upon the previous order."[3] While America's involvement in World War I had fueled discussion about whether Western civilization was declining, there had been little concern over the possibility of global annihilation. The detonation of the atomic bomb at the close of World War II shattered such equanimity. Moreover, specific and continuing policy considerations over the next decades prevented such concern from dissipating into abstract discussion; the development and possible deployment of the more destructive H-bomb, nuclear testing, and the ensuing arms race fueled debate within scientific and popular communities and remained on the agenda of the now political "responsible scientists."[4]

"Along with admiration and exaggerated expectations," writes historian Paul Boyer, "the atomic bomb also brought expressions of deep apprehension and hostility toward science— expressions that became more insistent as the postwar glow of pride gave way to sober second thoughts."[5] Although public opinion polls in 1949 revealed that, by and large, Americans were more concerned with national security than they were frightened by fallout from nuclear testing, polls from the previous year had indicated that an "anti-scientific sentiment" was growing in the country.[6]

On one level, the bomb simply provided the latest manifestation of technology that "humankind must control," a recurring sentiment that expressed a long-standing ambivalence toward technology. Indeed, the language and attitude of many critics of the bomb, chiefly religious critics, echoed technology's critics both before the bomb was detonated and afterward. "We are Frankensteins," lamented one postwar preacher, "who have created a technological civilization that in the hands of sin can literally exterminate us. Unless great ethical religion can catch up . . . our science will be used to destroy us."[7]

The bomb also saw the development of an outlook that could properly be called "antiscientific." In 1949 the editor of *Scientific American* opined that society now regarded scientists "with almost hysterical fear." "To the average civilized man of 1950," he concluded, "science no longer means primarily the promise of a more abundant life; it means the atomic bomb."[8] After World War II members of the scientific and medical establishments themselves began calling publicly for the control of technology.

The highest profile and most admired critics of atomic policy came from among those scientists who had worked to develop the bomb and who came to feel that, after August 6, 1945, "it was their urgent duty to try to shape official policy regarding atomic energy."[9] Not otherwise politically radical, these "atomic scientists," in trying to control scientific research policy, were characterized by the ambivalence toward progress that has been a continuing historic feature of segments within America's intelligentsia.

In December 1945 scientists concerned with the course of science in a post-atomic world launched the *Bulletin of the Atomic Scientists.* Two of its board members, Eugene Rabinowitch and Morton Grodzins, affirmed that the *Bulletin* had two aims: one was to address fellow scientists about the connection between science and politics; the other was to address the public at large about the "many dangerous presents from Pandora's box."

> One purpose was to make fellow scientists aware of the new relationships between their own world of science and the world of national and international politics. A second was to help the public understand what nuclear energy and its application to war meant for mankind. It was anticipated that the atom bomb would be only the first of many dangerous presents from Pandora's box of modern science. Consequently it was clear that the education of man to the realities of the scientific age would be a long, sustained effort.

Securing international control of nuclear energy was a practical goal of the scientists, an objective in which they ultimately failed. While the immediate trigger for their political activism was the devastating implications of atomic warfare, the authors and editors of the *Bulletin* meant to attend to issues beyond the atom. They also were concerned more broadly with "the impact of science on man and society and the appropriate responses to that impact, and to the relations between science and other spiritual and material forces which mold and reflect the minds of men: that is, religion, ethics, law and art."[10]

The political awareness and activism of the atomic scientists did not mark the first time that scientists had come to understand the political and social implications of their research. It was, however, the first time that scientists with the celebrity of the atomic scientists had made their case in so public a fashion. Their influence was extensive, in part, because of the public's fascination with those who had played such a crucial role in constructing "a doomsday weapon that had killed more than a hundred thousand human beings."[11] And it was not simply the "public" per se that reacted so profoundly to atomic developments; geneticists, who viewed discoveries in their own field as explosive, found in the atomic scientists' experience a cautionary tale, a warning they chose to heed.

Bioethical caution toward medical and scientific advance grew out of concerns first expressed by the postwar bioscientific community. Among those scientists influenced by the atomic science movement was a small but influential group of international geneticists concerned with what they believed to be a "biological revolution." When the claims and fears of these geneticists became public, some intellectuals in the United States became alarmed. This alarm helped to call forth the growth of bioethics.

Genetic discoveries during the 1950s and 1960s were interpreted by their founders as unprecedented and morally challenging. The geneticists believed that the public must be made aware of their implications. Referencing the atomic experience in their calls for interdisciplinary scrutiny of biomedical research and development, geneticists followed consciously in the footsteps of the postwar "responsible science movement," which had called for greater thoughtfulness about the regulation of atomic power. It was an expanded interpretation of "the biological revolution" that elicited public response and helped to provoke bioethics into existence. Although the world had been unprepared for the atomic age, it was incumbent upon scientists, geneticists believed, to prepare the world for the biological revolution, especially for its eugenic implications. Should this responsibility be ignored, geneticists would not only be morally culpable but could ultimately lose control over the course of their research as well. While some geneticists were worried about the eugenic implications of their work, others sought to implement eugenic practices. The umbrella issue in either case, though, was how to deal with the public, whether it be to educate or to manage.

Eventually, public intellectuals, including those who later would be considered bioethicists, began to respond to calls from scientists for external assistance. How these intellectuals would parlay the difference between educating and representing the public, on the one hand, and managing the public, on the other, became one of the great silent social developments of the decade.

Biologist Jean Rostand's work in parthenogenesis and artificial insemination revolutionized those fields. In his more popular 1959 work, *Can Man Be Modified?*, Rostand explained that biologists-geneticists had "great fear" that atomic energy had caused atomic mutations. As a result of these mutations, "science ha[d] declared war on the hereditary patrimony of mankind."

> We can produce *artificial mutations* and cause new races of living creatures to appear on earth. It is indeed ... upon these experiments with artificial mutation—the first of which were due to Muller in 1927—that the great fear felt by biologists in face of the liberation of atomic energy is based, and based very solidly: they know, in fact, that a fresh cause of mutations has been introduced by this into the world; and, knowing also that almost all mutations—if not all—tend to deteriorate the species, they find themselves forced to conclude that ... science has declared war on the hereditary patrimony of mankind.[12]

Biologist Julian Huxley corroborated this fear over the effects of atomic radiation on human genetics at a 1962 Ciba Foundation conference, where he offered his reasons for believing that the human population was deteriorating genetically: "The evidence is mainly deductive," he explained, "based on the fact that we are preserving many more genetically defective people than before, and are getting a lot of radioactive fallout."[13]

At the 1962 Ciba Foundation conference, geneticists formally discussed various long-standing issues,[14] declared the existence of a biological revolution, and pondered how to manage

this revolution vis-à-vis the public. A conference held the following year at Ohio Wesleyan University revealed another
strain of concerns. Whereas several of the Ciba geneticists wished
to influence the public in order to advance their favored eugenic programs, the Wesleyan geneticists sought to warn the
public of the eugenic implications of genetic research.

At the 1962 meeting the Ciba Foundation sponsored a symposium that was, in the words of its director, Gordon Wolstenholme, "an exceptional conference." Putting aside the usual
technical aspects of medical research, the conferees directed
their attention instead to the social implications of current biological research.[15] Wolstenholme stressed the particular influence of the atomic research precedent in prompting this
concern; and he urged that the potential for research findings
to alter human life in drastic ways required the attention of
every human being. Rather than emphasize the threat of mutations, as Rostand had before him, he stressed the need for preparedness—a social, political, and ethical preparedness that
had been absent before nuclear power.

> The world was unprepared socially, politically, and ethically for
> the advent of nuclear power. Now biological research is in a fer-
> ment, creating and promising methods of interference with "nat-
> ural processes" which could destroy or could transform nearly
> every aspect of human life which we value.
>
> Urgently, it is necessary for men and women of every race and
> color and creed, every intelligent individual of our world, to con-
> sider the present and imminent possibilities. They must be pre-
> pared to defend what they hold good for themselves and their
> neighbors.[16]

Wolstenholme's concern was raised by the attitudes toward eugenics expressed by some of his geneticist colleagues, including the views presented at the Ciba conference by scientists
Hermann J. Muller and Joshua Lederberg.

In 1946 Hermann Muller was awarded the Nobel Prize in
Physiology and Medicine for the "discovery of the production
of mutations by means of X-rays."[17] His presentation at the 1962

conference was not about such a technical issue but, rather, about the social problem posed by "genetic deterioration." "Society now comes effectively to the aid," he admonished, "of those who for whatever reason, environmental or genetic, are physically, mentally, or morally weaker than average." Owing to improved standards of medicine and living in industrialized countries, the percentage of those born who do not reach maturity has shrunk—a situation, he suggested, that might contribute to genetic deterioration and to an excess of "genetically defective adults." Compounding this problem, Muller noted, is the inclination of "persons possessed of greater foresight . . . and keener regard for their family" to have few children, whereas "those who are clumsier, slacker, less provident, and less thoughtful are the very ones most likely to fail in keeping the number of their children down." Because he despaired of influencing the decision to reproduce of those he considered to be genetically degraded, Muller suggested instead that society concentrate "not on the number of children in a family but on their genetic composition." This would be achieved by what he called "germinal choice"—namely, creating more possibilities for people to opt for artificial insemination from vetted semen. For Muller, "preservation will allow the accumulation of larger, more diverse stores, their better appraisal, and the fading away of some of the personal biases and entanglements that might be associated with the donors."[18]

Thus, he advocated as a eugenic tool a biomedical technology that had been developed to treat infertility. Muller felt that artificial insemination to improve the human gene pool should not be imposed but should be offered as a matter of choice. Therefore, it was important, he believed, to work toward changing public mores in order to advance the first "empirical steps" toward germinal choice.

At the same conference, Stanford University's Joshua Lederberg, a fellow Nobel Prize–winning geneticist, offered his version of these eugenic predilections. Most geneticists, he generalized, are "deeply concerned over the status and prospects of

the human genotype." "Do we not still sinfully waste a treasure of knowledge," he asked rhetorically, "by ignoring the
creative possibilities of genetic improvement?" Whereas Muller
suggested promoting germinal choice, Lederberg's pet proposal was *euphenics*, a term he defined briefly as "the engineering of human development": "*Development* is the trans-
lation of the genetic instructions of the egg, embodied in its
DNA, which direct the unfolding of its substance to form the
living, breathing organism." Although his discussion of euphenics is vague, Lederberg's desire to utilize recent genetic
findings to manipulate genes and "improve" mankind is unmistakable. "It would be incredible," he argued, "if we did not
soon have the basis of developmental engineering technique to
regulate . . . the size of the human brain by prenatal or early
postnatal intervention. . . . Needless to say, 'brain size' and 'intelligence' should be read as euphemisms for whatever each of
us projects as the ideal of human personality." After scant discussion of some developments in medical and scientific research, Lederberg casually referred to "the medical revolution" (i.e., another variant of the "biological revolution") and
urged that this revolution "should begin to arouse anxieties
over its orderly progress."[19]

The eugenic inclinations were unmistakable and unabashed.
Perhaps the most extreme eugenic posture at the Ciba conference was that of Nobel laureate (and codiscoverer of DNA)
Francis Crick. He questioned whether having children should
be considered a right at all and entertained the policy of allowing people to reproduce only after being licensed to do so
in order to discourage the "genetically unfavorable" from con-
ceiving.[20] These eugenic proclivities did not go unchallenged.
Economist Colin Clark likened what he was hearing to Nazi
genocidal doctrines and accused the "brilliant and misguided
scientists" of beginning a second cycle of eugenic doctrines.[21]

The following year, in April 1963, geneticists met again at
the Ohio Wesleyan University conference. Once again, the influence of the atomic experience was unmistakable. Also notable was the perceived need to anticipate the social problems

their research might precipitate—something that, geneticists felt, the atomic physicists sadly had neglected. Unlike the Ciba geneticists, however, who were concerned chiefly with promoting their various eugenic programs, several of the Ohio Wesleyan conference participants hoped to educate and warn the public about the potential social effects of genetic research.

The comments of conferee Dr. Salvador Luria explained that the responsible science movement was influencing geneticists both to seek useful applications of scientific developments and to avoid "evil applications."

> The impact of science on human affairs imposes on its practitioners an inescapable responsibility. This responsibility actually affects the course of scientific development: on the one hand, it created the urge to seek useful applications and to foster their general acceptance; on the other hand, it may restrain the scientist from pursuing a line of research that is clearly leading to evil applications. The instance of nuclear fission research is a natural illustration of the many moral alternatives that face the natural scientist in his work.

Luria professed that scientists had a "special responsibility," namely, "that of informing the public of the actual and potential applications of their findings and of the possible consequences." He added that "most thoughtful physicists would agree that they could have done more to anticipate the almost inevitable developments of nuclear physics and to inform society of the impending challenge."[22]

Luria's colleagues revealed similar influences and concerns. Expressly invoking the comparison "between the position in biology now and the position in physics about 1935," Dr. Guido Pontecorvo cautioned that the analogy was "not quite correct," and implied that scientists had an even greater duty now to keep the public apprised of developments than there had been before.

> By about 1935 the release of nuclear energy was theoretically certain. What was very doubtful in the minds of practically all physicists was whether it would ever have any practical interest at all.

Because of this uncertainty, and the then-prevailing idea that scientists should mind their own business, physicists did next to nothing to inform society of what might come. When they realized that practical applications were only a matter of effort, the war was on and, with it, secrecy. The result was that the momentous decisions, first of making the bomb and then dropping it, were made by a handful of men inevitably not expressing public opinion on this matter. . . .

In biology today the position is almost the reverse. On the one hand we are convinced that, one way or another, profound advances in the technology of human engineering are going to come and could be speeded up by adequate concentration of effort. On the other hand, we have no theoretical grounds for predicting which particular approaches ... are more likely to be workable. Finally, biologists, and in general all scientists, today have learned from the experience of nuclear energy and are conscious that it is their duty to inform society of the implications of the advances in their own fields.[23]

Dr. Rollin Hotchkiss also compared developments in genetics with those of atomic science. Unlike the political impetus for atomic weapons, he anticipated disturbing commercial ramifications for genetic manipulation.

The impetus (for genetic manipulation), unlike that which developed ATOMIC WEAPONS, will not be political, but rather commercial. In a country where, during every waking moment, one is being told to acquire and enjoy the products of industrial ingenuity, we can well expect that one will be told he owes it to himself to improve his own genes, as well as his neighbor's! And governments will not want to stand in the way of such initiative—unless informed thinking on the part of the public is strongly unfavorable to it.[24]

The geneticists' particular urgency stemmed, in part, from their belief that genetic research findings constituted a biological revolution. "Biology has undergone a revolution," cautioned geneticist T. M. Sonneborn in his preface to the edited record of the symposium, "the scope and impact of which not even the penetrating imagination of Aldous Huxley could sense. . . . New possibilities of controlling human development have emerged."[25] Pontecorvo offered a historic perspective: "Human engineering in one form or another has," he advised, "been practiced on a minor scale for a long time. Prenatal care, immunization, blood transfusion, organ transplantation, plastic surgery, supplementation . . . of the diet are all practices of human engineering that we have taken in our stride." This more subdued viewpoint notwithstanding, even Pontecorvo believed that there was great need for circumspection, education, and the development of ethical standards "and democratic means." Failure to attain these goals would send us down the path to Aldous Huxley's brave new dystopia.

> The ethical issues raised by these past feats of human engineering are qualitatively no different from those we shall have to face in the future. The difference will be quantitative: in scale and rate. Even so, the individual steps may still go on being so small that none of them singly will bring those issues forcibly to light: but the sum total is likely to be tremendous. That is why we have to look for those issues now . . . and make them a matter of universal knowledge. Only this way can we hope to develop ethical standards and democratic means. . . . The alternative is a wise oligarchy knowing and doing what they think good for the rest in the way so vividly pictured by Aldous Huxley 40 years ago in *Brave New World*.[26]

Affected by the responsible science movement and the experience of the atomic physicists, scientists in other fields endeavored to alert the public to the existence of the biological revolution and to entreat others, from outside science, to assist in the creation of ethical standards to manage this revolution. The calls from such scientists were met by critiques that emerged in the incendiary cultural environment of the sixties.

The far more radical critiques of the 1960s raised the stakes of ethical considerations of science and technology. These critiques considered not merely the aims and assumptions of research scientists narrowly construed, but broader evaluations of how science and society interacted and buttressed each other in limiting both individual freedom and social justice. Lewis Mumford, Herbert Marcuse, Jacques Ellul, and Theodore Roszak offered evaluations not of specific technologies but of "the technological society."[27] Their analyses helped to inject a far more critical rejoinder to the geneticists and scientists who had originally stimulated such evaluation.

Spanning several decades, the prodigious commentary of social critic Lewis Mumford demonstrates something of the radical transformation that characterizes post-atomic cultural life. In his 1934 *Technics and Civilization*, Mumford offered an optimistic discussion of the ages of history: the eotechnic, the paleotechnic, and the neotechnic.[28] Roughly coterminous with what other commentators referred to as the second industrial revolution of the 1920s, Mumford's neotechnic era, driven principally by electric power, was to rid society of the ills prevalent during the coal-driven paleotechnic age. The early Mumford described a social and technological trajectory of improvement. By the 1970s, however, his characterization of technology and society had shifted, as early sanguineness gave way to disillusionment.

The turning point for Mumford was the 1945 bombing of Hiroshima and Nagasaki, acts which he found abhorrent and which left him shocked and dismayed. "The news of these events," according to historian of technology Thomas Hughes, "so stunned him [Mumford] that discourse became impossible for him for days afterwards."[29] Mumford's anguish over the creation and use of the bomb pervades volume two of *The Myth of the Machine*, which he titled, *The Pentagon of Power*. Reminiscent of intellectual predecessors and anticipating the bioethicists who would come to be his contemporaries, Mumford emphasized humankind's need to take control of technology: "If the key to the past few centuries has been 'Mechanization Takes Command,' the theme of the present book may be summed up in Colonel John Glenn's words on returning from orbit to earth: 'Let Man Take Over.'"[30]

Mumford departs from his merely ambivalent predecessors,

however, when he goes beyond the simple and vague prescription that "mankind" needs to assume control. He identifies the source of technological drive and the entity that must be controlled, indeed, curtailed: the megamachine.

> The megamachine ... is *not* a mere administrative organization: it is a machine in the orthodox technical sense, as a "combination of resistant bodies" so organized as to perform standardized motions and repetitive work. But note: all these forms of power, one re-enforcing the other, became essential to the new Pentagon of Power.[31]

The bomb, according to Mumford, was pivotal in the creation of the modern megamachine:

> The production of the atom bomb was in fact crucial to the building of the new megamachine, little though anyone at the time had that larger objective in mind. For it was the success of this project that gave the scientists a central place in the new power complex and resulted eventually in the invention of many other instruments that have rounded out and universalized the system of control first established to meet only the exigencies of war.[32]

The organization and stealth that were necessary to create and deploy the atomic bomb led Mumford to discover the modern megamachine. With the identification of the megamachine as the driving force behind "mankind's" technological progress, Mumford moved beyond ambivalence and toward the more critical, more radical posture symbolic of the sixties. In this aspect, Mumford is also distinct from the bioethicists, who continued along within the tradition of ambivalence. Mumford marks the modern bifurcation of approaches to technological development—between critiques wary of the effects on society of certain technologies, on the one hand, and more radical analyses that view technological production as part of larger, more insidious institutional and cultural imperatives, on the other.

The menace of atomic devastation was also a concern for Herbert Marcuse in his 1964 work, *One Dimensional Man*. "Does not the threat of an atomic catastrophe," he began, "which could wipe out the human race also serve to protect the very forces which perpetuate this danger? The efforts to prevent such a catastrophe overshadow the search for its potential causes in contemporary industrial society."[33] This popular work, however, was less a direct response to the atomic explosions (as the responsible science movement had been) than a thoroughgoing indictment of "advanced industrial society." In dramatic prose, Marcuse laid out the case that individuals were being dominated by the forces of technological production, a powerful and thorough domination of which few were aware.

The technological society, Marcuse argued, by manufacturing false needs and feeding all material desires, choked out every impetus for change, creating "unfreedom." Hiding behind claims made to an "objective rationality"—"a technological veil"—the forces of exploitation had become as invisible as they were pervasive:

> Within the vast hierarchy of executive and managerial boards extending far beyond the individual establishment into the scientific laboratory and research institute, the national government and national purpose, the tangible source of exploitation disappeared behind the facade of objective rationality. Hatred and frustration are deprived of their specific target, and the technological veil conceals the reproduction of inequality and enslavement. With technical progress as its instrument, unfreedom—in the sense of man's subjection to his productive apparatus—is perpetuated and intensified. . . . decisions over life and death, over personal and national security are made at places over which the individuals have no control.[34]

Marcuse's words were a call to arms. Critics ambivalent about the social and cultural effects of technology merely called for caution. Marcuse exhorted his readers to make "the Great Refusal." Whereas Mumford maintained a degree of optimism, Marcuse was largely pessimistic about any possibility for significant change or ultimate freedom. Mumford maintained his optimism in that he believed it possible to escape from the hold of the megamachine through individual awareness and dissent.

If I dare to foresee a promising future other than that which the technocrats (the power elite) have been confidently extrapolating, it is because I have found by personal experience that it is far easier to detach oneself from the system and to make a selective use of its facilities than the promoters of the Affluent Society would have their docile subjects believe.

Though no immediate and complete escape from the ongoing power system is possible, least of all through mass violence, the changes that will restore autonomy and initiative to the human person all lie within the province of each soul, once it is roused. Nothing could be more damaging to the myth of the machine, and to the dehumanized social order it has brought into existence, than a steady withdrawal of interest, a slowing down of tempo, a stoppage of senseless routines and mindless acts. And has not all this in fact begun to happen?[35]

Mumford believed, nevertheless, that what little hope for change there was had to come from a complete rejection of the entire system.

Like Marcuse, Jacques Ellul's critique, read widely in American intellectual circles, was not simply a call for cautionary oversight of technology. Society, for Ellul, was overpowered by what he called "technique." Technique included machine technology but was far more than this. "In our technological society," he enjoined in his 1964 work, *The Technological Society*, "*technique* is the *totality of methods rationally arrived at and having absolute efficiency* . . . in *every* field of human activity." It is a process responsible not only for all scientific, medical, and technological output, but for all politics and all economics as well. As technique (the rational means to any end) grows, the individual becomes alienated from the ends or purpose of work, becomes less human, less spontaneous, less spiritual— all is rationalized, calculated, planned: "No human activity is possible except as it is mediated and censored by the technical medium. This is the great law of the technical society."[36] There is, for Ellul, no hope of change because technique is inexorable.

Theodore Roszak, like Ellul before him, wrote of the dehumanizing technological, rational processes of social, scientific, political, and economic institutions. Drawing explicitly on Ellul's work, Roszak affirmed the pervasive role of "the technocracy." Roszak's technocracy was "that social form in which an industrial society reaches the peak of its organizational integration. It is the ideal men usually have in mind when they speak of modernizing, up-dating, rationalizing, planning." The technocracy was run by experts and by those who employ the experts. Those who govern in a technocratic society "justify themselves by appeal to technical experts who, in turn, justify themselves by appeal to scientific forms of knowledge. And beyond science, there is no appeal." The technocratic process was as subtle as it was pervasive. Even those in the state and corporate structure who perpetrated it were unaware of their functionary role. Technocracy was a "grand cultural imperative" that "devours the surrounding culture," a totalitarianism "perfected because its techniques become progressively more subliminal."[37]

Writing at the end of the decade and affected by the revolutionary potential of the counterculture that he described, Roszak offered a more hopeful view than did Marcuse or Ellul, of the possibilities for change from a dispiriting social system to one fit for humanity. Like the antimodernists of the nineteenth century, the countercultural youth of the 1960s sought escape from technocratic totalitarianism through alternative life-styles. The scientific world view had to be subverted, and the "young centaurs" of the counterculture were best suited to subvert it. It was not unprecedented, Roszak noted, to have such an antirationalist force like the counterculture in our midst. "What *is* new," Roszak claimed for the large numbers of middle-class youth he characterized, "is that a radical rejection of science and technological values should appear so close to the center of our society."[38]

The social critiques of Mumford, Marcuse, Ellul, Roszak, and others functioned, in part, as rallying cries for a serious challenge to existing institutions, infusing important arenas of the intellectual culture of the 1960s. Throughout the sixties, scientists continued to alert the public to a need for thought-fulness about the products of scientific research. "Science marches on, fast and furiously," cautioned biophysicist Leroy Augenstein in 1969, "but all too often our ability to handle our newfound powers does not keep pace." As Augenstein saw it, scientific developments were forcing scientists to play God. There was no other option.

> Science is literally forcing us to play God. And let's make no mis-take about it, we do this to a certain extent all the time. . . . The big change is that the ante in this poker game had gone sky-high. . . .
>
> Must we play God deliberately and extensively by making such awesome decisions? The answer is a definite yes. In fact, we no longer have any option as whether we will or will not "play God."[39]

Such concerned scientists were hailing from within Amer-ica's historic tradition of ambivalence toward technological de-velopment. Certain lay intellectuals, however, who began re-sponding to the requests of "responsible scientists" for public awareness called out from a different cultural environment. They began agitating in the brackish intellectual waters be-tween traditional American ambivalence toward progress and the more combative cultural milieu fast becoming the decade's cultural banner.

By the late 1960s, some intellectuals—some of whom would eventually don the appellation *bioethicist*—began heed-ing scientists' call to participate in working through the ethi-cal and social questions provoked by the biological revolution.[40] "Geneticists are so confident of being able to tamper with heredity," wrote Gordon Rattray Taylor in *The Biological Time Bomb*, "that they have begun to warn us to beware of them."[41] Indeed, after opening the Pandora's box of public input, ge-neticists were unable to control the tenor of open discourse. The emerging commentary was often not merely advisory but critical and, at times, hostile. For some public intellectuals and science popularizers working in the radically charged post-atomic milieu—individuals like Taylor, Amitai Etzioni, and Donald Fleming—ambivalence became antagonism.

Taylor's 1968 best-selling *Biological Time Bomb* was translated into some twenty languages and read widely around the world.[42] His popularization of the new genetics demonstrated the important social legacy contributed by the atomic bomb's detonation. Taylor also offered dramatic evidence, as did other popularizers, that the bomb was chief exemplar of science run amok. Like others before and after, Taylor effectively invoked that icon of science without ethics, Dr. Frankenstein. "The explosion of the first atom bomb drove a jagged crack through the superman image [of the scientist]," he explained. "From behind the mask of the beneficent father-figure, the mad engineer suddenly looked out, grinning like a maniac. As the impact of the biological time-bomb begins to be felt, the haunted look of Dr. Frankenstein may gradually appear on the faces of the biologists." His forecast implicated the emergence of an "anti-scientism."

> Science will begin to be seen in a very disenchanted way, as the bringer of gifts which too often end by canceling their own benefits.
>
> . . . There may well arise a solid opposition to science, an anti-scientism the extremer elements in which may demand the prohibition of all scientific activity except under special license and direct supervision by non-scientific representatives of the state.[43]

The tone of Taylor's analysis exceeded the cautionary tenor typical of milder critiques. His presentation was an antipathetic challenge to scientific authority. The target of attack was not simply humanity's inability to cope with too rapid technological advancement. His evaluation implied a disaffection with scientists themselves. Although some biologists had published warnings, he explained, and a "few imaginative scientists" had attempted to examine the implications of their work, there were too few scientists so engaged. Moreover, he observed, the warnings were published in journals too specialized to be generally accessible. Critiques like Taylor's made clear that the geneticists who had announced the biological revolution—and who had either tried to steer it in a eugenics direction or who had asked for assistance in dealing with the eugenic implications of their work—would not be able to control fully the nature or direction of the examination for which they had called.

Sociologist Amitai Etzioni's article for the popular periodical *Science* was similarly disturbing in its explanation of the effects on society of certain reproductive technologies. In "Sex Control, Science, and Society," Etzioni's critique of the adjustments required of society by new biological technologies echoed the sensibility of heedfulness expressed by generations before him, by peoples who were merely ambivalent toward progress. Society will be overwhelmed if the pace of change is not slowed.

> We are aware that single innovations may literally blow up societies or civilizations; we must also realize that the rate of social changes required by the accelerating stream of technological innovations . . . may supersede the rate at which society can absorb. Could we not regulate to some extent the pace and impact of the technological inputs and select among them without, by every such act, killing the goose that lays the golden egg?

But Etzioni went further. If the scientific community did not accept its responsibility by examining the effects of its research findings on society, it may be unable to protect itself from a "societal backlash and the heavy hands of external regulation."[44]

In his 1969 article for the *Atlantic Monthly*, "On Living in a Biological Revolution," historian Donald Fleming introduced the reading public to the biological revolution and its revolutionaries. His cautionary tone not only revealed his fears about the new science but also implicated the scientists themselves. "Their ideal is the manufacture of man," he warned. "This is the program of the new biologists—control of numbers by foolproof contraception; gene manipulation and substitution; surgical and biochemical intervention in the embryonic and neonatal phases; organ transplants or replacements at will." "The great question becomes," he warned, "what is it going to be like to be living in a world where such things are coming true?" His answer was a solemn forecast of a "general acquiescence" to a chilling type of human perfection.

> The will to cooperate in being made biologically perfect is likely to take the place in the hierarchy of values that used to be occupied by being humbly submissive to spiritual counselors chastising the sinner for his own salvation. The new form of spiritual sloth will be not to want to be bodily perfect and genetically improved. The new avarice will be to cherish our miserable hoard of genes and favor the children that resemble us.[45]

Public intellectuals responded to the scientists' calls for caution with furious alarm. The radically infused culture of the sixties had transformed the recurring historical concern over humankind's inability to cope with technoscientific advance into a sharp-edged, perils-focused combat with the creators of bioscientific "progress."

Against a cultural and intellectual tradition of ambivalence embodied in the worries of the postwar atomic science movement, and of the contemporary hostility of intellectuals toward science and technology, bioethics emerged in bold relief. Bioethicists differed from their predecessors in the tradition of ambivalence in that they brought to bear specific modes of reasoning and philosophical principles on a wide range of specific medical and technological issues.[46] Either a "teleological" or a "deontological" ethics, for example, may be employed to examine the principles of autonomy and beneficence as they arose in issues surrounding the physician-patient relationship, human experimentation, death and dying, abortion, genetic research, and the like.[47] Bioethics was a more issue-specific, technology-by-technology enterprise than historic ambivalence toward technoscientific advance. Bioethicists differed from their radical contemporaries by focusing singularly on the products of biomedicine, thus resisting claims made to larger connections between science and society.

Bioethicists took the mantle of the biological revolution offered up by the geneticists beginning in the 1950s and broadcast by public intellectuals in the late 1960s; and in the early 1970s, they cast this veil over other biomedical research and procedures that were provoking concern, including organ transplantation, kidney dialysis, amniocentesis, recombinant DNA, mechanical ventilation, and the redefinition of death. The biological revolution focused their attention, fired their imagination, became their raison d'être. It became also, for them, an inexorable process. The overarching bioethical posture (as with historic ambivalence) took the medical, scientific, or technological product as a given. Their largest consideration devolved not on the interconnectedness between science, society, and the technocratic megamachine, but on the more ethereal— and, therefore, less politically threatening—reflection about how humankind could come to terms with the inescapable results of biomedical research.

The self-reflective writing of bioethicists, as opposed to their considerations of specific ethical issues, taken together with their social location among the educated classes, highlights the position of bioethics within the tradition of irresolution toward technological development. The sine qua non of the early bioethical explanations, for example, was to express concern over either "modern biology," the "biological revolution," or modern technological advances.

The 1971 opening statement of the first issue of the *Hastings Center Report*, a periodical dedicated to the discussion of bioethical issues, revealed bioethicists' anxiety over technology: "The goal of this Report ... is to advance public and professional understanding of the social and ethical problems arising out of advances in the life sciences."[48] Other bioethicists followed suit, frequently imparting a sense of urgency accented by a perception that there was something qualitatively different about recent scientific advances. For example, former vice-president Walter Mondale (an early bioethics supporter) argued in the *Hastings Center Report* that "recent advances in biology and medicine make it increasingly clear that we are rapidly acquiring greater powers to modify and perhaps control the capacities and activities of men by direct intervention into and manipulation of their bodies and minds."[49]

Throughout the 1970s and 1980s concern over the biological revolution remained paramount. Bioethicist Robert Veatch, for example, cautioned that, "the biological revolution has given great new powers to medicine. . . . Medical ethical problems that once were no more than entertaining speculations about the future are now a reality."[50] For George Kieffer, writing a "textbook of issues" for bioethics, "Scientific knowledge, including that coming from biological study, is the root of technological changes which are transforming our social and political environment. We have to decide how to deal with them or the decisions will be made by default."[51] And as bioethicist H. Tristram Engelhardt explained, "Philosophical reflections have been directed to health care because of . . . major and rapid technological changes that have created pressures to reexamine the underlying assumptions of established practices."[52]

Founders of the field sometimes cited other grounds for the emergence of bioethics.[53] Additionally, with the passage of several decades since the emergence of bioethics, some bioethicists, such as Arthur Caplan, have become more cautious about claims made for the primacy of biotechnology. His remarks both acknowledge bioethics' traditional focus on technology and register a disaffection with this technological determinism:

> We may not feel that bioethics should be driven by technologies or only try to grapple with the way some newfangled gizmo or gadget works, but it's clear that the public face of bioethics has been driven by transplants and artificial devices and making babies in test tubes. . . . The theme that emerges throughout all these stories is that ethics is struggling to keep up with technology. No matter how hard you think and no matter what you do, you're trailing behind Dr. Edwards, Dr. Jarvik, Dr. Bailey, or Dr. Starzl.[54]

But whatever other reasons were given, and despite more recent qualifications, the sense of needing to come to terms with a new biology and technology was, initially, of preeminent concern.[55]

Joseph Fletcher's "prebioethical" work, *Morals and Medicine*, was an ethical analysis of a number of issues including sterilization and euthanasia.[56] The book, written in 1954, "fell upon deaf ears," according to Daniel Callahan, president of the Hastings Center, the nation's first bioethics institute. The reason, Callahan argued, was that most of the technological advances associated with bioethics had not yet "taken hold." Then in the 1960s, "great changes" began to occur. Prior to World War II the federal government had invested large sums in biomedical research. The "payoff" began, Callahan explained, in the 1960s: "That decade saw the advent of organ transplants, the widespread use of the respirator, and the emergence of issues concerning the allocation of resources. The question of the definition of death and the care of the dying also appeared on the horizon. Prenatal diagnosis and genetic screening were in an experimental phase." Callahan relates how bewildering these developments seemed at the time. "By 1969 the issues were becoming identified and were obviously important," he writes. "Yet there was still an air of science fiction about them. They had not quite taken hold within the medical establishment, and the media by and large tended to treat them as exotic problems on the horizon rather than as problems that would soon face day-to-day medicine."[57] Callahan's explanation suggests how bioethics' stock-in-trade came to be its identification of (and implied strategies for solving) moral quagmires brought about by biomedical advances.

Ironically, bioethics' singular focus on the ethical dilemmas engendered by new biotechnologies ultimately defused the post-atomic fury of mistrust that had helped call public attention to bioscientific and biomedical developments in the first place. By taking production of exotic biotechnologies and procedures as inexorable, bioethics truncated the political indictments of radical critiques that had emphasized interconnections between society, science, and the technocratic "megamachine." In this way, biotechnological producers, so vilified by sixties culture, escaped the potentially debilitating scrutiny that had helped launch bioethics as a social phenomenon.

Because the chief contemporary explanations of bioethics tend to focus on the 1960s, they miss the importance of the formative nature of post-atomic culture of the 1950s as well as the historically recurring national ambivalence toward development.[58]

The most recent interpretation of the rise of bioethics is University of Washington bioethicist Albert Jonsen's *The Birth of Bioethics*. Jonsen declares that "the seeds of bioethics were sown ... during the forty years between 1947 and 1987.... Those years show the medical and biological sciences advancing with astonishing rapidity and record many questions about such wondrous advances." Jonsen chooses 1947, the year of the Nuremberg Trials and the promulgation of the Nuremberg Code, as "a new beginning in the moral traditions of medicine, a beginning that would become bioethics."[59] But the shocking nature of the Nazi atrocities that prompted these events stands apart from what Jonsen himself and other bioethicists took as the hallmark of bioethics, namely, the many questions about the "wondrous advances" that were developing with "astonishing rapidity."

Although Jonsen's chronology includes the years that cradled the post-atomic movement, he leaves largely unexamined how the horrors associated with the development and deployment of the atomic bomb fueled self-reflexive fears among scientists about science and about scientists themselves—fears that went on to ignite alarm among public intellectuals writing in the radicalized culture of the 1960s. In fact, the anxieties of the earliest bioethicists and "prebioethicists" seem to have focused more acutely on the atomic bomb than on the atrocities committed by Nazi doctors on human subjects, although ethicists would later occasionally invoke the latter by way of example. The high-profile brutality of Nazi scientists doubtless affected both popular ethical awareness and professional sensibilities. But the bomb sullied popular images of scientific purity and, consequently, threatened to attenuate scientific autonomy in ways that Nazi experimentation could not. The bomb was a rueful challenge to the self-understanding of the "responsible scientist," and responsible scientists conveyed this self-doubt to the public. While deeply disturbing, Nazi experimentation ultimately could be understood as part of the enemy's greater villainy. It was not possible to distance Allied science from the bomb in this way; atomic devastation was the sorry, inescapable, product of "neutral" science.

Jonsen acknowledges that medical researchers in America made a distinction at the time between medicine and Nazi medical atrocities. "The lesson of Nuremberg," he writes, "seems to have made little impression on the world of medical research." He cites historian David Rothman, who notes that "The prevailing view was that [Nuremberg medical defendants] were Nazis first and last; by definition nothing they did and no code drawn up in response to them was relevant to the United States." For "blunter" effect he quotes Dr. Jay Katz: "It was a good code for barbarians but an unnecessary code for ordinary physicians."[60] But if American medical researchers themselves were not unduly concerned about how the atrocities of the Nazi doctors might affect or inform their practice in the 1940s, how did this barbarism trigger the "beginning that would become bioethics" decades later (when bioethicists and bioethical institutions come into existence)? Jonsen's analysis does not really bridge this gulf. His methodology seems to have been to take issues identified by bioethicists in the 1960s, and then to look at the individual chronologies of those issues (research on human subjects, genetics, organ transplantation, issues surrounding death and dying, and human reproduction). This task is different from asking why and how bioethics developed historically and came to regard those issues as problematic—something that the radical cultural diffusion of post-atomic fears and America's recurring ambivalence toward development does explain.

The interpretations of University of Pennsylvania sociologist Renee Fox and Columbia University historian David Rothman explicitly pinpoint the decade of the sixties in their explanations of the development and nature of bioethics. Fox focuses on and problematizes the technological determinism of bioethics (the view that biomedical advances constitute the driving force behind the development of bioethics). Rothman's chief focus is on a different bioethical hallmark, namely, the emergence of unparalleled challenges to professional authority and the posing of difficult ethical questions. But they both center their analyses on the bioethical activity of the late 1960s, implicitly concluding that bioethics emerged out of that decade's agitated reconsideration of power and authority.

Renee Fox had been one of the original members of the Hastings Center (initially called Institute for Society, Ethics and the Life Sciences, or ISELS). Despite this early association, however, she became critical of the emerging field. Her more censorious observations earned Fox the whimsical title of "bioethics basher" from some of her erstwhile colleagues.[61] Fox did not produce a historical analysis of bioethics per se; but it is possible to cull a historical location of the phenomenon from her characterizations more generally. In her 1989 essay on "The Sociology of Bioethics," Fox invoked but did not analyze the broader social and cultural environment in her description of bioethics' genesis. "Bioethics . . . surfaced in American society in the late 1960s," she explained, "a period of acute social and cultural ferment." Her central view of the emergence of bioethics, however, is that it was a response (a critical, one-sided response, she insists) to scientific and medical technological developments—not the dynamic product of social turbulence. For Fox, bioethics had "concentrated its attention on a particular group of advances in biology and medicine. With strikingly little acknowledgment of the improvements in identifying, controlling, and treating disease that these advances represent, bioethics has focused on actual and impending problems that medical scientific progress has brought in its wake."[62]

Fox described what she called "the ethos of bioethics." This ethos, preoccupied chiefly with truth telling, distributive justice, and the "principle of beneficence," is also technologically deterministic, she argued, because fresh focus on these concerns was made necessary by the challenge of biomedical advances. From its inception in the 1960s up to the 1980s, according to Fox, bioethics went through three overlapping stages. Despite her acknowledgment that bioethics grew out of a concern with the ethical dilemmas associated with various technologies, Fox's account of these stages is not about technological advances per se. Instead, she characterizes three overlapping interests: experimentation on human subjects; death, dying, and the nature of personhood; and issues relating to the economics of health care. She generalizes that, "Along with the thematic shifts through which bioethics has passed, there have been fluctuations in the prominence accorded to particular medical scientific advances and types of medical treatment."[63] Her account of bioethics, however, does not analyze the technologies and biomedical advances that she claims are the chief bioethical concerns.

Most important, the ethos of bioethics was, according to Fox, structured within and limited by an individualist framework. The firm bioethical conviction that individualism (individual rights, autonomy, and self-determination) is a nonnegotiable virtue led Fox to one of her chief criticisms of bioethics, namely, its lack of concern with social problems.

> The skein of relationships of which the individual is a part, the sociomoral importance of the interdependence of persons, and of reciprocity, solidarity, and community between them, have been overshadowed by the insistence on the autonomy of self as the highest moral good. Social and cultural factors have been primarily seen as external constraints that limit individuals.

Shaped by professionals, scholars, and academics, bioethics is, according to Fox, conservative in important ways. Its technological determinism (which ignores social and cultural influences) and its individualist predilection "bend it away from involvement in social problems." Thus, when bioethicists focused on nontreatment decisions in neonatal intensive care units, for example, "relatively little attention [is] paid to the fact that a disproportionately high number of the extremely premature, very low birth weight infants . . . are born to poor, disadvantaged mothers, many of whom are single nonwhite teenagers."[64] Bioethics "reduces" social problems to fit within a framework that is utilitarian, positivist, and reductionist.

Whereas Fox was critical of bioethicists, historian David Rothman was critical of Fox. He found her "guilty of some of the very charges she leveled at bioethics." If bioethicists failed to place the movement in a societal framework, Fox also failed to do so, he argued, "leaving the impression that bioethicists were a self-seeking and self-promoting group of academic entrepreneurs." Whereas Fox saw bioethics' focus on individualism as reductionist and conservative, Rothman saw the bioethical commitment to individual rights as "the core of its success"—something that explained what, he insisted, Fox did not explain, namely, bioethics' broad appeal. And if the professional and academic backgrounds of bioethicists made them uncomfortable with more overtly "left-leaning" social advocates, the "conceptual similarities" bioethicists shared with such advocates were critical: they all "looked at the world from the vantage point of the objects of authority, not the wielders of authority."[65]

The lack of political understanding of class, race, and gender that Fox found so limiting was, for Rothman's view of bioethics, an advantage. Whereas Fox labeled the bioethical ethos to be reductionist and apolitical, Rothman viewed the same outlook as part of bioethics' durability and widespread appeal; as he put it, "bioethics crossed class lines":

> [Bioethics] was at least as responsive, and perhaps even more so, to the concerns of the haves than the have-nots. Not everyone is

poor or a member of a minority group or disadvantaged socially and economically, but everyone potentially, if not already, is a patient. This fact gives a special character and appeal to a movement that approaches the exercise of medical authority from the patient's point of view.[66]

Rothman's narrative of the emergence of bioethics did not construe the concern with technology as a defining characteristic: "To simply assume that bioethics is the product of changing technology ignores all the interesting questions and doesn't do justice to the complexity of the situation. The technology itself rarely sets off the problem. It's how we feel we have to respond to the technology that ultimately gets interesting."[67] Instead, his analysis focused squarely on the antiauthoritarian, antiprofessional posture of bioethicists in challenging medical hegemony. Prior to the 1960s, medical ethics had largely been the "doctors' preserve," Rothman believed. Established medicine insisted that "medical ethics should be left entirely to medicine, and whatever public policies flowed from these ethical principles were not to be contested or subverted." In the 1960s a new set of questions developed over issues surrounding experimentation with human subjects and revolutionized this state of affairs. For Rothman, "human experimentation served as the magnet bringing outsiders to medicine."[68]

The public and bioethical response to revelations of ethical abuses in experimentation with human subjects in the 1960s accounts for what Rothman believed to the hallmark of the origins of bioethics, that is, bringing public critical examination to bear on medical decision making. This scrutiny, Rothman asserted, came from many directions, from many professions:

> The ethics of human experimentation became one of the first issues to bring an extraordinary array of lawyers, philosophers, clergymen, and elected officials into the world of medicine. Thus, the events in and around 1966 [when a number of ethical abuses were revealed] accomplished what the Nuremberg trials had not: to bring the ethics of medical experimentation into the public domain and to make apparent the consequences of leaving decisions

about clinical research exclusively to the individual investigator.[69]

The important development, as he saw it, is that once violations were revealed, medical research came under "collective surveillance."

But by focusing chiefly on developments within experimentation on human subjects, Rothman paid insufficient attention to the fact that researchers in the areas of genetics, organ transplantation, and the redefinition of death also were calling for ethical assistance from sources outside of the profession—a phenomenon that suggests that other social factors were operating.[70] Moreover, Rothman's fixation on the antiauthoritarian aspect of bioethics caused him to undervalue another major bioethical characteristic—a characteristic that Fox had highlighted but did not fully evaluate, namely, its technological determinism.

Rothman did not ignore the fact that anxiety over technology was a major focus of the era. He accepted unproblematically, however, claims made regarding the special nature of the technology in question, even while he made an effort to place this technology within a social context. "In the 1960s," he noted, "medical procedures and technologies ... posed questions that appeared to some physicians—and to even more non-physicians—to go beyond the fundamental principles of medical ethics or the expertise of the doctor and to require societal intervention." Modifying an interpretation that bioethicists themselves expounded, Rothman stressed not the technology, which he unquestioningly agreed was presenting unprecedented ethical quandaries, but the special nature of the times. This technology "appeared at a special time," he insisted, "when Americans' long-lived romance with machines was weakening"; the "test case" of this technology—organ transplantation—"engendered a whole series of novel dilemmas." For Rothman, the more important aspect of this period is not so much that this technology gave pause. It was, rather, that "an entirely new group of people [bioethicists] ... were ready to pose the questions and even attempt to answer them."[71]

The development of bioethics, Rothman argued, saw "an enormous amount of confrontation." At the 1992 Birth of Bioethics conference, Rothman highlighted this conflict and explicated the stakes:

> I do not think this is a peaceable story. . . . I think there was a lot of tension; there was a lot of conflict. And the reason for that conflict is not difficult to understand. The stakes were very, very high. One was talking about professional sovereignty to the profession that enjoyed the maximum amount of sovereignty.
>
> By no stretch of the imagination did bioethics easily make its way . . . into medicine.

Rothman then designated the impetus for this bioethical challenge—the civil rights ideology:

> The point I want to make is that ideology was terribly important. . . . The ideology of the [bioethics] movement in its first context—and I'm more interested in a movement than a field—had to do with civil rights. . . . What it meant was that the patient was understood as a member of a minority. It was as though the patient was a tenant in a housing project, as though the patient was—these models, obviously, with all the feminist language intended—a woman; as though the patient was in some very, very particular way on welfare, powerless.[72]

For Fox, bioethics' individual-rights framework for assessing the ethical implications of advances in technology led bioethics away from a more thoroughgoing consideration of social problems. For Rothman, bioethics' challenge to professional authority placed it squarely within the fray of popular defiance of social institutions. Both Fox and Rothman, however, implicitly assume that bioethics is a novel product of the 1960s. Yet this assumption is problematic.

The cultural environment—fashioned, in part, by events like the Greensboro sit-in of 1960, the Port Huron statement in 1962, the Free Speech Movement of 1964, and the Democratic convention of 1968—was, of course, a manifestation of defiant social critiques. The dominant image of such events and the legacy of such critiques make it too tempting to assess all social phenomena from the era as coming from or responding to the cultural idiom of the 1960s. The temptation is especially great because sixties critiques often linked social injustices to technocratic structures and mind-sets. Free Speech leader Mario Savio's emblematic exhortation to the crowd on Berkeley's Sproul Plaza in 1964, for example, implicated a technocratic menace, "the machine." "There's a time when the operation of the machine becomes so odious," he urged them, "makes you so sick at heart, that you can't take part, you can't even tacitly take part. And you've got to put your bodies upon the gears, and upon the wheels . . . and you've got to make it stop. And you've got to indicate to the people that run it, the people who own it, that unless you're free, the machine will be prevented from working at all."[73]

Although bioethical idiom and discourse were undeniably infused by this emblematic radical culture, characteristic bioethical concerns over the effects of bioscientific discovery were part of a larger recurring social phenomenon. Bioethicists, as philosophers, theologians, lawyers, doctors, and biologists, fit better within the hoary history of an educated American elite discomfited by the less agreeable social and cultural consequences of technological production. Seen from this broader vantage, the unique development in the 1960s is not that critics challenged America's technological imperative. The distinctive features of the 1960s are that members of the lay intelligentsia adopted critiques that medical and scientific professionals themselves had begun calling for during the 1950s, and that lay-bioethical oversight became institutionalized. In this sense, bioethics can be understood as an expansion of the postwar responsible science movement.

Rothman's account of bioethics, wedded to viewing bioethics as part of the larger challenges to authority of the 1960s, largely ignored how many calls for public oversight of medical and scientific research were raised by scientific and medical professionals. Although he noted a number of different revelations of ethical violations, Rothman focused chiefly on Dr. Henry Beecher's 1966 publication "Ethics and Clinical Research" in the *New England Journal of Medicine*. Here Beecher details twenty-two studies of experimentation with human subjects involving ethical violations.[74] As Rothman viewed it, the important development was that once violations were revealed, medical research came under "collective surveillance." He acknowledged, but regarded unproblematically, the fact that Beecher was himself a physician, making his exposé the work of a whistle-blowing "insider."

Indicative of its roots in the responsible science movement, physicians and scientists were an important part of the earliest bioethics institutes.[75] Later, as legal ramifications of biomedical developments became clearer, lawyers also became a part of this interdisciplinary cadre.[76] The original core of bioethical literature, however, was produced by philosophers and theologians.[77] These commentators who eventually would be known as bioethicists (who, as revealed in the notes and texts of their works, had read the public intellectuals' alarming accounts and knew of the concerns raised by the post-atomic geneticists) negotiated the field between ambivalence and hostility variously. The work of Episcopalian ethicist Joseph Fletcher and that of Methodist minister Paul Ramsey demonstrate the spectrum of bioethical views.

Joseph Fletcher is often considered the first bioethicist. His 1954 *Morals and Medicine*, however, is more a Protestant theologian's rejoinder to the Catholic tradition of medical ethics (as he termed it, "a dialectic discussion with Catholic opinion") than a consideration of the biological revolution. Fletcher distinguished himself from the Catholic tradition (and from most other bioethicists) by celebrating the increased choices that biomedical progress offered humankind, rather than emphasizing the problematic features of such advance. Fletcher's analysis, moreover, was patient-centered: "The patient is not a problem; he is a person with a problem." And the patient had

rights: the right to know the truth, the right to control parenthood, the right to overcome childlessness, the right to foreclose
parenthood, the right to die. Fletcher's work was a caveat for
physicians: "If doctors undertake to care for the health of our
people, they must undertake to do it conscientiously. This is to
say that they have to put their consciences to work, as they do
their skills." But, his work was not an attack on science or
America's scientific imperative—the cast of mind that radical intellectuals regarded with such foreboding. "Science,"
Fletcher opined, "in spite of its frequent tragic misuses, con-
tributes to our moral range and the magnitude of our ethical
life; technology not only changes culture, it adds to our moral
stature."[78]

Six years after its first publication, the 1960 edition of
Morals and Medicine had taken on a new urgency. Fletcher
now underscored the impact of recent medical developments.
"Medicine is making tremendous and exciting advances," he
explained, "and these advances constantly provide new depth
and new circumstances for the moral discussion." Imparting a
sense of exigency, Fletcher instructed that "we are already
launched into the 'post-modern' era; and we must learn to live
in it, not only technically, but ethically." Still, although his
work took on the anxious quality of those concerned with the
dawning biological revolution, Fletcher remained celebratory
of scientific advance and the choices it offered humankind.
The making of these choices, Fletcher believed, made "man"
truly human, allowing him to emancipate himself from nature's "brutal existence."[79]

By contrast, Paul Ramsey regarded both physicians and bio-
science itself with an air of challenge. In *The Patient as Person*, Ramsey adopted Fletcher's posture toward doctors and
stressed the importance of valorizing the rights, dignity, and
basic humanity of patients as people. He urged that "the doctor makes decisions as an expert but also as a man among men"
who must be "attentive to the patient as a person."[80] Ramsey
went beyond Fletcher, however, and questioned the authority

not merely of practitioners but of biomedicine as an unchallenged enterprise.

In *Fabricated Man*, his 1970 rejoinder to those post-atomic geneticists who sought to promulgate a new eugenics, Ramsey clearly laid out the difference between ethicists with a "frivolous conscience" who only rationalize whatever science makes possible and those of a "serious conscience" who are willing to take the position that some things should never be done.

> We need to raise the ethical questions with a serious and not a frivolous conscience. A man of frivolous conscience announces that there are ethical quandaries ahead that we must urgently consider before the future catches up with us. By this he often means that we need to devise a new ethics that will provide the rationalization for doing in the future what men are bound to do because of new actions and interventions science will have made possible. In contrast, a man of serious conscience means to say in raising urgent ethical questions that there may be some things that men should never do. The good things that men do can be made complete only by the things they refuse to do.[81]

The larger philosophical difference between Fletcher and Ramsey yielded differences with respect to the virtue of specific biotechnologies. Whereas Fletcher embraced artificial insemination as "morally lawful," for example, Ramsey felt that artificial insemination with donor along with other biotechnological possibilities "debiologized" people. "When the transmission of life has been debiologized," he explained, "human parenthood as a created covenant of life is placed under massive assault and men and women will no longer be who they are." Fletcher believed that bioscientific discovery brought more choices and, with those choices, increased possibilities for rational decision making—a cogitative act that marked the boundary between animals and human beings. Ramsey, by contrast, was less concerned with demarcating the line between the animal and human species and more concerned with not overstepping the boundary between the human and the divine: "Men ought not to play God before they learn to be men, and after they have learned to be men they will not play God."[82]

Bioethicists in the 1990s remained as anxious over how to negotiate these boundaries and over the sociopolitical ramifications of this negotiation as they were in the late sixties and seventies. In September 1992 pioneering bioethicists gathered to celebrate thirty years of bioethics at the Birth of Bioethics conference in Seattle, Washington.[83] In his report on the conference, bioethicist Albert Jonsen (borrowing Callahan's phraseology) related the concern over the role of bioethics as "a friendly, not hostile force within medicine."

> Had the critical edge that Ramsey honed in *The Patient as Person*, warning about the moral dangers of unreflective acceptance of medical advance, become only the moral enthusiasm about medical progress manifested in the thinking of Joseph Fletcher? George Annas wondered whether bioethics has been coopted by medicine. Has bioethics, in Rothman's words "rather heroic" in its early days, slipped into, as Arthur Caplan colorfully phrased it, "sway-bellied middle age, looking pretty institutional, with its people talking ex cathedra and flexing their expertise.[84]

This concern over co-optation contrasts with the more sanguine views of Willard Gaylin, psychiatrist, bioethicist and cofounder of the Hastings Center. Gaylin advised in 1990 that society should not be "victimized by the Frankenstein factor," that is, overvaluing the dangers of "high-technology research" and fearing any research that holds the possibility of altering the human species. Although he is uneasy over technology's role in causing society to feel "progressively impotent in the face of pleasureless social institutions that we ourselves created but that now seem to control us," he urged, finally, that "no technology to date has been able to dehumanize and demoralize with the power of drugs, poverty, neglect, despair, narcissism, and blind hedonism."[85]

While bioethicists were and remain conflicted over what their social function should be, institutional forces have chosen their role for them. Ultimately, although the rancorous cultural environment of the 1960s buoyed early efforts of bioethicists at institutionalization, afforded bioethicists a special cultural currency, and even inclined some of them to adopt a tone of indignation toward authority, the dominant bioethical posture has become not one of antagonism toward medicine and science as players in a larger field of cultural domination; rather, it is one of modest ethical discussion about the results of medical and scientific research. Institutionalized in this position by the mid-1970s, bioethics—wittingly or unwittingly—helped to transform the era from the radically charged, confrontational cultural milieu of the 1960s to an age in which ethical experts manage problems generated by a technology that is seen, ironically, as value-neutral in its creation, even while it is problem-causing in its outcomes. Functioning in this manner was the price to be paid for institutional longevity. How this came to pass is considered in the next chapter's examination of the Hastings Center, the world's first, and premier, bioethics institution.

Chapter Two

"Leader of Leaders"

THE HASTINGS CENTER, 1969 TO THE PRESENT

In April 1986 the American College of Physicians bestowed its Loveland Award on the Hastings Center, to acknowledge the center's "distinguished contribution in the health field." According to the college, the Hastings Center was the "conscience of medicine, psychology and philosophy."[1] In 1991 the National Institutes of Health, confirming its trust in this "conscience," awarded the Hastings Center a grant to explore the ethical aspects of the Human Genome Project.[2] Never before had a federally funded science project specifically designated 5 percent of its budget to pursue an ethical examination of its work.[3] Here I examine the nation's first and premier bioethics institute, the Hastings Center, and, in so doing, explore the nature of the conscience to which the government entrusted the ethical evaluation of the human genome.

The history of the Hastings Center is a story of the challenges facing pioneering bioethicists in their struggle to ensure institutional longevity while safeguarding autonomy and integrity. In the 1960s and 1970s, physicians and scientific researchers sought external, interdisciplinary advice, sometimes out of a genuine quest for answers to ethical questions surrounding new technology, sometimes to protect against liability while pursuing technological innovation (see Chapter 3). But while such counsel often was welcomed, excessive criticism would not be tolerated. Indeed, by 1975 members of the fledgling bioethics community would describe an "ethics backlash" against their efforts, waged by medical and scientific institutions.

The Hastings Center found it had to chart the narrow strait

between backlash and support by adopting a "nonideological" posture and by shying away from even the perception of being an "activist" organization. Whatever the actual content of members' beliefs, the center's self-promotional efforts stressed its role as mediator, translating between contending lines of thought rather than as advocate, directly challenging medical or scientific authority.

Moreover, the Hastings Center did not, in the final analysis, cast the need for examining medicoscientific behavior principally in terms of power imbalances between doctor and patient or between science and society—positions typical of critiques of the professions during this period (see Chapter 1). Instead, Hastings Center members construed their role as coming to grips with the social implications of the biological revolution.[4] Analyzing a biological revolution came to mean studying the effects of a preexisting technology on society, not an examination of the social production of that technology. It was, and remains, an effort to grapple with a disturbing array of ethical dilemmas generated by technologies that are seen as value-neutral in their creation even while they are problem-causing in their outcomes. The conservative position into which the center fell was not foreseen at its inception, however. In the watershed moment of the late 1960s and early 1970s, some bioethicists tried to exploit real but frail options made possible by the radicalized culture of the era; they were largely frustrated in these efforts, however, by established biomedical interests.

Eventually, by construing issues as ethical rather than political the Hastings Center (and bioethics more generally) secured absorption into existing medical and scientific institutional frameworks. But achieving longevity in this way meant that the intense, sweeping examination of political values undergirding medicine and science—an analysis emblematic of 1960s commentary and practice—would slip away.[5] The waxing of the social discourse of ethics tells one of the stories of the waning of the sixties. Declawed, social critique moved from protest to management.

The center began operating in New York in 1969. Briefly

dubbed "Center for the Study of Value and the Sciences of Man," the group changed its name two more times: to "Institute of Society, Ethics and the Life Sciences" (ISELS) in 1970, and finally to "the Hastings Center."[6] Apart from its unique status as the world's first bioethics organization, the Hastings Center deserves special consideration for a number of reasons: it is the only "independent" bioethics organization (i.e., unaffiliated with any university, medical school, or law school); its publications include the influential *Hastings Center Report* and *IRB: A Review of Human Subjects Research;* and many of the current leaders in the bioethics movement are now or have been "fellows" or associates of the institute. Among those once affiliated with the Hastings Center who now head bioethics programs elsewhere are George Annas, Boston University; Arthur Caplan, University of Pennsylvania; Alexander Capron, University of Southern California; and Robert Veatch, Kennedy Institute of Ethics, Georgetown University.

The center's primary goal was to bring an interdisciplinary approach to bear on problems engendered by biomedical advance. In 1975, the *Hastings Center Report* began including a section that explained the center's function and purpose:

> The Institute was founded in 1969 to fill the need for sustained, professional investigation of the ethical impact of the biological revolution. Remarkable advances are being made in organ transplantation, human experimentation, prenatal diagnosis of genetic disease, the prolongation of life and control of human behavior—and each advance has posed difficult problems requiring that scientific knowledge be matched by ethical insight.

In trying to cope with the wide range of ethical, social, and legal questions, the institute established three general goals: advancement of research on the issues, stimulation of universities and professional schools to include ethical examination as a part of their curriculums, and public education.[7] Originally, the areas under consideration were death and dying, genetics, behavior modification, population control, experimentation on human subjects, and the ethics of social science investigation.[8]

Additionally, the Hastings Center held summer workshops and initiated projects designed to stimulate the teaching of biomedical ethics in universities and professional schools around the country.[9] The center's organizational structure consisted of a president, a director, a board of directors, a permanent staff, and a group of approximately one hundred nonresident "fellows"—the philosophers, theologians, psychiatrists, lawyers, physicians, and biomedical researchers located around the country who offered their talent to examine research projects through research "task forces."[10] The center also sponsored student interns and postdoctoral scholars. In 1996 it could boast an associate membership of twelve thousand.[11]

Eventually, twelve to fifteen fellows came to meet about four times a year for two or three days to present papers and participate in discussions on designated research topics. Participating fellows would vary from year to year as different specialists matched up with changing research agendas.[12] They might be called on often to report on conclusions and render opinions before congressional hearings. "When we take up a problem," related Hastings Center founder Daniel Callahan in 1979, "we start at the theoretical level, but we carry it through to some practical application."[13]

It became typical for research groups to suggest guidelines for the ethical use of a new biotechnology. When genetic tests were developed to screen for sickle cell anemia in the late 1960s, for example, the Hastings Center's genetic group recommended that these tests be voluntary and confidential and that post-test counseling be made available. These guidelines, designed, in part, to protect African Americans (who are the group most commonly affected by this disease) from insurance discrimination, were published in the *New England Journal of Medicine*.[14] Similarly, a number of states adopted the center's model for the 1968 redefinition of death. The minutes from most of these various task force groups are, unfortunately, not a part of the Hastings Center archives. Still extant at the archives, however, are the minutes, memos, and letters regarding issues discussed at fellows and board of directors meetings.

These records offer a glimpse of the dilemmas, successes, enthusiasm, and frustration experienced by first-generation bioethicists on the road to institutionalization.

Suggesting some of the early exuberance felt by the original participants, Callahan tells how many of his colleagues willingly absorbed the costs of the meetings and early work: "a good deal of support was being gained from various professionals and others. They were willing to pay their own way to early meetings and do a good deal of work without any financial return and were enthusiastic about developing a solid core of people in the country who would take on these ethical problems in a systematic way."[15] The earliest financial support, other than the absorption of these personal costs, was a loan in March 1969 from Callahan's mother. By the end of that year, the fledgling institute had received grants from Elizabeth K. Dollard and John D. Rockefeller III, along with a matching grant from the National Endowment for the Humanities and the Rockefeller Foundation. Concerned with maintaining its autonomy, the founders eschewed both university affiliation and corporate sponsorship. Unable to do without the corporate sponsorship it sought to avoid, however, the institute eventually accepted this type of support to supplement the individual contributions, private and federal grants, membership dues, and income from educational programs. The center's budget was $1 million by 1977 and $1.3 million by 1981. In the early 1980s the center began an endowment campaign.[16] In February 1996 the center reported a budget of $2.3 million.[17]

The official Hastings Center story regarding its genesis begins with a Christmas party in 1968, an event and its sequelae which Callahan has related in a third-person narrative.

> During the latter part of 1968, Callahan talked with a number of people about the idea of forming an organization devoted to ethical problems in the life sciences. It was not clear exactly what was needed, and there were no models to draw upon. At a Christmas party in 1968, Callahan mentioned the idea to a Hastings neighbor, Dr. Willard Gaylin, a psychiatrist and psychoanalyst on the faculty of the College of Physicians and Surgeons, Columbia Uni-

versity. Gaylin was taken with the idea, and over the next couple of months they talked intensively. Each then set about talking further with various friends and colleagues. Having been trained in philosophy, Callahan knew a number of potentially interested people in the theological and philosophical communities, and Gaylin had a variety of contacts in the medical and biological communities. In March of 1969 . . . they organized a small meeting of people interested in the idea of forming a research and educational organization. That, probably, was the public beginning, and for well over a year the center was located in Callahan's house (with a mimeograph machine in the home of Will Gaylin).[18]

"A Preliminary Sketch," written in 1969, is Callahan's earliest statement making the case for creating a bioethics institute. It expresses the need for an organization to provide sustained analysis of dilemmas posed by scientific "advance." Although he does not use the term *biological revolution,* the notion is so "fundamental" as to "require no persuasion":

> Those of you to whom I am sending this preliminary sketch require no persuasion, or background information, on one fundamental point. As a result of scientific, technological and social developments, mankind is now faced with a wide, pressing and almost bewildering range of ethical problems. Some are old and some are new. Either way, however, the response made to these problems will have much to do with the future of mankind.
>
> Few deny the existence of a fundamental unrest about the deepest issues of meaning and value. Few deny that some of the most agonizing problems of our time are basically ethical. Few deny that they must be confronted. Yet what do we find when we ask some pointed questions? How well are these problems being met?[19]

A statement the following year invoked the "biological revolution" explicitly:

> The Institute is a response to the growing recognition that advances in the life sciences pose social and ethical questions touching the fate of individuals and societies, now and in the future. The common phrases "biological revolution," "population explosion," and "environmental crisis" only hint at the terrible complexity of these advances and the problems that follow in their wake.[20]

While the center's bioethicists were captivated by the idea of a biological revolution, they were not unaffected by the radical critiques of science and society then permeating the culture of the 1960s. In fact, the center came to fruition after more than a decade of radical social critique of medicine and science (see Chapter 1). A number of the center's members participated in this discourse of challenge. In a 1996 article, Renee Fox, reflecting on this early period, characterized bioethics as being "conservative" from the very beginning: "Unlike Dan Callahan, I would not portray the initial years of U.S. bioethics as a period of moral debate and 'nay-saying' that shook the foundations of mainstream thinking.... It was in the area of human experimentation that American bioethics had the greatest reformist impact at its inception.... And yet, from the outset, the Geist of bioethics has been markedly conservative."[21] In distinguishing herself from Callahan, however, Fox demonstrates the divergent ways in which bioethicists recall their past. The archival record suggests that, although the Hastings Center did settle into a conservative posture by the mid 1970s, this posture was not a foregone conclusion from the outset; it was the ultimate result of a fluid situation in which radical cultural influences created real but fragile possibilities— possibilities some bioethicists tried to pursue.

Daniel Callahan, for example, relates that "early on" he had been "caught up" in the critiques of technology and society then pervading popular and academic culture—the works of Jacques Ellul, Herbert Marcuse, and Theodore Roszak among others.[22] From this broader concern with "how technology changes the context of human life," Callahan came to focus on ethical issues surrounding medical technologies. Bioethics, he believes, was affected by the "anti-establishment criticisms of experts" inhering in the notion of "patients' rights." In a 1977 interview Callahan revealed this challenging posture: "Doctors want ... to make all the choices. Well, we're saying to them, no. There are some public interests at stake here and some general principles you have to abide by.... You're playing in a public ball-park now ... and you've got to live by certain standards ...

like it or not."[23] Not simply a comment regarding the need to be circumspect about the cultural effects of technological development, Callahan's remarks resonate with the political challenge to authority characteristic of the more radical sixties critiques. Callahan's 1991 oral history reflected on the early years of bioethics and laid out three factors that led to its development: the burst of new technologies that presented unprecedented moral dilemmas, a sixties culture-of-rights that demanded a more democratic participation in the doctor-patient relationship, and the decade's characteristic penchant for creating new organizations, of which bioethics was one.[24]

The work of bioethicist and Methodist theologian Paul Ramsey represents the small but influential segment of scholarly output that reflected the sixties culture of challenging established authority, institutions, and values. In his 1970 treatise, *The Patient as Person,* Ramsey not only questioned the wisdom of the trajectory of biomedical science but also underscored the erosion of scientific authority by impugning the once privileged province of scientific expertise: "Galloping technology," he admonished, "gives all mankind reason to ask how much longer we can go on assuming that what can be done has to be done or should be. . . . These questions are now completely in the public forum, no longer the province of scientific experts alone."[25]

Similarly, in "The Generalization of Expertise," Robert Veatch (later president of the Kennedy Institute of Ethics in Washington, D.C.) emphasized the need for limitations on scientific expertise and prerogative: "Generalization of expertise arises when . . . it is assumed that an individual with scientific expertise in a particular area also has expertise in the value judgments necessary to make policy recommendations simply because he has scientific expertise. This assumption is pervasive in decision making in scientific areas, but unwarranted."[26]

But if Hastings Center bioethicists were affected by general and critical evaluations of medical and scientific authority, the center gained legitimacy and endured despite this inclination. The center's survival rested on its ability to adapt to an emerging regulatory framework for the administration of professional medicine and science. The experience of the Hastings Center in securing legitimacy is the story of a kind of natural selection: the institutional environment surrounding medicine and science selected those characteristics of the fledgling ethics institute that were necessary for mutually assured survival— of both the bioethics community and the continued ideological trajectory of science and medicine.

The meetings and correspondence of the center's first year reflect the exuberance of individuals working in concert. Sociologist Renee Fox, one of the "Founding Fellows," wrote Callahan expressing some of this excitement: "I very much enjoyed our meeting, one of the purest, loftiest and most stimulating I've attended in years. The kind of grown-up idealism and communion that unites us is really rare."[27]

The early years that followed saw the rapid acceptance of the institute's authority by the press and other organizations. The founders themselves were awed at their fast-paced success. In the summer of 1969 Daniel Callahan stalled a reporter from *Time* magazine, who wanted to make reference to the institute in an article on ethics and biology. Its practice only just begun, Callahan felt unsure whether the institute was "ready yet to see some publicity."[28] By 1971, however, its professional reputation was secure, as was an agreeable and enduring relationship with the press:

> Perhaps the most surprising development of the past year ... has been the extent to which we have become known around the country. This does not mean that everyone can recite the exact name of our Institute ... but under one name or another people know we exist. We receive an average of 10 inquiries a day, requesting information on the Institute itself, or one of our programs. The greatest stimulus for this has come from a number of news stories about the Institute (in *Time*, the *New York Times*, *BioScience*, the newsletter of the Ford Foundation and the National Endowment of the Humanities ...).[29]

But while the institute's fellows had clear reason for satisfaction on this score, they were less secure on other fronts. Perhaps most poignantly, founders grappled in earnest with the insecurity of not knowing whether they could, in fact, devise answers to the questions for which the press, Congress, professional medicine and science, and the "public" so urgently sought solutions. As the field grew and bioethicists became more "professional," the pressure to find resolutions to ethical dilemmas became more acute.

By 1974 the Hastings Center had achieved stability after four years of growth, which saw a doubling of activities and budget each year. Indeed, the field of bioethics was growing more generally. In the early seventies, the National Endowment for the Humanities made $2.5 million worth of grants to the various bioethics groups that had organized to study biomedical ethics.[30] Reflecting on these years, Callahan expressed some of the agonies associated with "professionalization" in his 1974 "Report to the Fellows." Pushed to have answers to dilemmas, the ethical "expert" often buries private doubt and uncertainty.

It's all such a goddamn hard mess, with wisdom scanty, knowledge slight, and terror just around the corner in one's mind. I can sit through three years of the meeting of our group on death and dying and learn to talk with some articulateness and precision about the end of life. But as soon as some friend or relative presents me with a real case, their case, I am reduced to fumbling and muttering, an enormous sense of inadequacy. What happened to all that expertise, that fast talk, those lovely, refined distractions? It's just gone, seeming to have nothing whatever to do with dying, really dying.

. . . Unfortunately, the role of professional (particularly the *paid* role) offers no special encouragement to indulge the gap between private agony and public statements. . . . One gets pushed, and tacitly approves the pushing, to stay away from the dark corners of doubt and uncertainty, of deeply felt inadequacy. One is asked to bring light, lucidity and logic, and one is only too happy to accommodate the request. They don't want a dolt, a muttering victim of private struggles with darkness and ignorance; they want a pro. . . .

... Somehow or other, as we become more professional, we will need to find ways of becoming less so. . . . I am convinced that the only way out is to keep returning to the discrepancy between how we tend to think about health and illness privately, late at night, and how we talk about them as professionals.[31]

When Hastings Center fellows held their annual meeting in 1974, Willard Gaylin and Daniel Callahan shared their concerns with colleagues. The institute was receiving calls continually, they told the fellows, "from legislators, educational institutions, parties to pending litigation, and others seeking its advice and assistance." Minutes from the meeting record that, according to Gaylin and Callahan, "the impact of the Institute is clear, but the officers are concerned about whether Institute personnel have the answers that people expect of them and whether their views are being regarded as too authoritative in an area in which definitive answers are not to be found."[32]

Moreover, behind the professional persona shown the public, the young "ethics institute" was beginning its struggle to find an ideological center. Although the institute explicitly eschewed adoption of any single political ideology in its earliest statements, operationally it proved impossible to avoid ideological issues entirely. Additionally, while its explicit goal was to offer criticism and advice on pressing social concerns, its unstated aim of obtaining cultural authority was susceptible to the larger, implicit interests of funding institutions. The debate over ideology focused on two master themes: whether to become an "activist" organization; and where to pitch itself on the prevalent pro-technology—antitechnology spectrum, a debate then fully engaged in wider intellectual and political circles.

The Hastings Center examined a variety of interpretations of the term *activist* in its earliest years. The meaning of the term could range from discussing ethical abuses and thinking about solutions, to devising guidelines, to seeking specific legislation, to calling press conferences in an effort to reveal abuses.[33] The kind of activism that the center found most nettlesome underscores the crux of the challenge then facing the founders: whether to respond to the increasing demand that they expose the ethical abuses they encountered, and whether they should do more to promulgate their own recommendations. Increasingly, they were called upon to address those critics who queried, "Why don't you get out of the ivory tower and into the streets?" and who complained, "You people should quit talking and get some new laws passed."[34]

The modus operandi of the Hastings Center became to form task forces on specific issues. These groups accepted grant monies for specific projects and, typically, created "guidelines" for proper (i.e., ethical) use of the applied technology in question. During its early years, however, the center was called on to go beyond this approach and to act decisively, in the manner of an advocate. The center's work in the areas of genetics and behavior control offers examples.

The center's genetics group published guidelines for mass genetic screening in the *New England Journal of Medicine.* Shortly thereafter, requests came forward for the institute to examine drafts of pending state legislation on screening programs. Additionally, and perhaps with greater potential force for social change, the institute was asked to join in a legal suit brought by a number of "black groups" against laws that allowed for mandatory screening for sickle cell anemia of schoolchildren.

The center's behavior control group was confronted with what it considered to be a similar "problem." In this case, a "legal-action center" based in Washington, D.C., was conducting an investigation of a Mississippi psychosurgeon who had used psychosurgery on young boys to "cure" their "disturbed condition" (records do not specify the nature of this condition). Members of the group were asked to join in the investigation, the ultimate goal of which was to bring suit against the psychosurgeon.[35] While such requests could be seen as opportunities for effecting social change, the center's founders viewed them as problematic and sought to discourage them.[36]

Another way in which center members avoided a more activist position was by deciding to refrain from making revela-

tions directly to the public. They chose, instead, to counsel professionals. In its annual "Report on Institute Activities" for
1972–73, it was clear that center members understood the "establishmentarian" bias of such a method:

> Our tendency to date has been to address ourselves mainly to pro
> fessionals rather than to the general public. We might, for in
> stance, have chosen to issue our guidelines on screening in the
> form of a public manifesto, complete with press conference, me
> dia coverage and all the other tactics used to publicize a message
> to the public. Instead, the genetics group chose to seek publica
> tion in a medical journal, aiming at those professionals who would
> actually be setting up and running screening programs. A similar
> decision was made by the death group, which also chose to pub
> lish its findings in professional journals. For some, of course,
> decisions of that kind are interpreted as "elitist," "establishmen
> tarian," etc.—at the very least as decisions to work within estab
> lished professional structures rather than to take the matter to the
> public.[37]

Two years later, the limitations of addressing professionals
rather than the public were still disturbing for some members.
Center fellow Robert Murray, for example, shared his concern
at the 1974 annual meeting of fellows. The fellows did not
bring their work sufficiently to the attention of the public, he
pressed. There was a danger of neglecting society at large as
distinct from specialists.[38] Despite reservations, however, the
methods of approach as well as the discourse became that of
the "ethics manager"—a style compatible with the growing
regulation of medicine, science, and technology of the seventies and eighties.

Callahan's reasons for preferring the nonactivist position
emphasized the desire to maintain collegiality among fellows
and avoid the "factionalizing" that would threaten the goal of
sharing different viewpoints:

> I do not believe that we should move in a much more "activist"
> direction, except under very special circumstances. In part, this is
> because I believe we are already doing effective, influential work,

which would not be particularly enhanced by a more activist focus. In part I also believe that, by serving as an available resource for groups and individuals pursuing particular abuses, we can make an effective contribution at that level on an informal basis. More generally, I think that any movement to take specific Institute stands on particular issues (as distinguished from, say, one of the research groups reaching some conclusion it wishes to publicize) would eventually (and quickly) lead to serious factionalizing within the Institute, making it exceedingly difficult for people of different ethical views and persuasions to continue working together in a mutually profitable way.[39]

He touched on another concern, however—the importance of maintaining a nonideological reputation:

Our public reputation might well change. At present I believe that we are looked upon as a non-ideological group, grinding no polemical axes, available as a trustworthy resource for all factions and ethical schools of thought. (This is not to deny that some observers, inside and outside, believe they can detect certain subliminal biases and predilections.)[40]

The perception that the Hastings Center was a "non-ideological group, grinding no polemical axes," was an image that worked well for the institute. It received favorable treatment not only at the hands of the press, but from funding institutions, including the Rockefeller Brothers, the National Endowment for the Humanities, and the Commonwealth Fund. Members of the U.S. Congress also affiliated their names with the institute, including then senator Walter Mondale. In June 1970 a letter from the office of Congressman William Steiger indicated some of the confidence the new ethics institute had engendered. Contacting the center for information about "the forthcoming genetics revolution," the letter's author remarked that, "If responsible public leaders do not undertake this task, I can foresee the day when some demagogue will come forward and run with this issue to the detriment of the nation and the scientific community."[41]

The "apolitical" aura of the center was carefully managed. The records reveal that, despite overt statements to the con-

trary, center fellows were struggling with a kind of technological skepticism in their efforts to come to terms with the biological revolution. Behind the image projected to funding
agencies and the press, institute members debated the wisdom
of their posture as a group of individuals with an "antitechnological" bias.

At a 1975 board of directors meeting, center officers reflected on their public image. "Ever since the Institute was organized in 1969," the minutes relate, "it has heard charges that
the perspectives, approaches and membership of the Institute
are (whatever we say about the matter) 'anti-science,' or 'anti-
medicine,' or 'anti-technology.'"[42] Indeed, nearly from its inception, the center wrestled with the reality and perception of
its ideological position.

After a meeting in 1971, Dr. Sheldon Wolff, then clinical director of the National Institute of Allergy and Infectious Diseases, wrote the center that he felt its members were "negative," had been "spinning wheels," and should have "spent
more time in trying to find out how society and the individual
might be improved."[43] Harold Green, one the fellows, defended the institute against Wolff's characterizations. Skepticism about biomedical improvements, he stressed, was an endemic part of their enterprise, one that prevented their work
from being a futile exercise:

> Although I am not an ethicist, it seems to me that our interest in
> ethics necessarily connotes a skepticism as to whether biomedical
> science and technology will improve society and the individ-
> ual. . . . We would be a sterile group indeed and would be engaged
> in an exercise in futility if we proceeded on the presumption that
> "biological research is not only important but vitally necessary for
> the continued improvement of our society." Even if we assume
> that this proposition is valid, there remains a substantial and le-
> gitimate question as to the appropriate scope and rate of publicly
> funded biological research.[44]

Similarly, the year prior to Dr. Wolff's admonitions, Willard
Gaylin, cofounder of the center, addressed one of the task forces
in a manner that reveals the ideological struggles of this "non-

ideological group." Gaylin tried to cajole the groups' members out of their skeptical inclination, as they undertook a study of the ethical implications of behavior control: "All of us share, perhaps justifiably," he urged them, "something of an anti-technological bias. . . . It would . . . be a useful exercise for us to consider some of the potential values of behavior control. . . . This exercise might, at the very least, enable us to explore our own bias a bit more carefully."[45]

Perhaps one of the clearest indications of the center's early struggles over ideology concerns the difficulty they had in inviting Joseph Fletcher to become a fellow of the institute. Author of *Morals and Medicine* and *Situation Ethics*, Fletcher was a prominent and somewhat controversial Episcopalian ethicist, a pioneer in the field of medical ethics, and well known for his celebration of what he considered the continuing benefits of science and technology (see Chapter 1). Significantly, he was overlooked for early fellowship. When, finally, the topic of his membership became an issue in 1970, the matter was handled with caution. Acknowledging that Fletcher was "a big name in the field" and that "on the whole, he should probably be asked," Gaylin and Callahan counseled the membership committee to proceed diplomatically as "some of our members intensely dislike him. . . . The main objection to him is that they don't much like his way of doing ethics."[46]

One of Fletcher's supporters commented that "the character of the group reflects the 'braking mechanism' on research more than the force which makes it go. Perhaps that is partly explained by the personalities and preferences of those who founded the Center."[47] Two years later Fletcher still had not been elected to fellowship. At that time, Gaylin characterized the institute, somewhat critically, as having drawn upon those who were less concerned with the benefits of science and technology than with their hazards; and he noted that Fletcher's absence as a fellow was, as such, no accident.[48]

In June 1972 members present at a board of directors meeting reached an informal conclusion: the institute should make an effort to elect more members with a pro-technology bias.

This casual resolution, made by less than a quorum, drew fire from Paul Ramsey. An esteemed and well-noted ethicist, Ramsey was known for holding views in marked contrast to Fletcher's pro-science celebration. Ramsey made clear his view that the board should limit itself to considerations of "matters of program, tasks to be undertaken, concern for a balance of disciplines (not *opinions!*) . . . , and the competence of persons . . . in their fields, their adherence to the purposes of the Institute, their openness to other disciplines, etc. (not whether they are *pro* or *con* 'progress' and all that!)."[49]

Three years into operation, why should the center have reconsidered criteria for composition of its membership and questioned the wisdom of its self-perceived antitechnological posture? The records reveal a specific concern that may have prompted a more general consideration of whether and how to allow all parties to express their views as they manifested themselves on the anti- versus pro-technology spectrum. In a push for consensus, published guidelines from the death group and the genetics group had not revealed the existence of dissenting viewpoints.[50] The issue was bothersome because the fact that the guidelines appeared to be unanimous was likely to have "had some weight" in their acceptance by the *Journal of the American Medical Association* and the *New England Journal of Medicine.*

> An issue worth serious discussion has emerged in some of the Institute research groups: whether and how the different groups should issue reports in the name of the whole group. The genetics group and the death group have both issued group reports (the former on mass screening and the latter on definition of death). In both instances there was an internal struggle, with some dissenters from the group consensus left rather unhappy that the consensus procedure excluded them. . . . The fact that the death group did develop a group statement on the definition of death problem no doubt had some weight in its acceptance by JAMA; and the same was probably true of the acceptance by the *New England Journal of Medicine* of the genetics group's statement on screening.[51]

The records also reveal, however, a subtler but perhaps more powerful pressure to come to terms with its technological stance—the importance of not alienating institutional sources of funding. At a meeting in the fall of 1971, Hastings Center fellows discussed funding and alluded to how the need for funds left them vulnerable to co-optation.

In October of that year, founding fellow Renee Fox addressed board members. She believed that the institute's growth was not simply a product of its work but a development that reflected change in American society. The issues they were considering had acquired social currency; they were now "legitimate." The danger attendant on this achievement, however, was the potential of their co-optation for "political ends." A memo to the board of directors summarized Fox's view, that vigilance was essential:

> Various groups in the future will probably begin struggling to make political capital of these issues. Of course the Institute should avoid letting itself be co-opted by others who would like to use either the Institute or the issues for political ends. The fact that the activities of the Institute are now very chic poses some dangers in itself . . . unless we are careful others may try to exploit us.[52]

The records reveal no further discussion of the co-optation issue as such. One of Fox's colleagues, however, raised a concern that was directly relevant: bioethicist and biochemist Leon Kass questioned whether too much of the center's work was initiated from outside the organization. Stressing that the institute controlled its own agenda, bioethicists Theodosius Dobzhansky and Willard Gaylin did not concur with Kass's apprehensions.[53] Certainly the center rejected studies that it felt compromised its sense of ethics. It would not, for example, help any group who wanted to use the center simply to learn how best to stave off legal liability. Two years after this meeting, however, Gaylin sent Callahan a memo indicating that he had, indeed, become aware of the very issue that disturbed Kass. Although the memo does not name the project in question, Gaylin's disillusionment is evident: "I ended up feeling," he lamented, "that here is a beautiful case of our taking on a project which we had not really desired to do, but have been sucked in a bit by the fact that there is a good deal of money available, and that they are interested in having us do it."[54]

Institutional independence was crucial from the beginning. Hastings Center founders had shunned the idea of seeking corporate funding. Fearful of corporate influence over their work, they preferred instead to seek monies from foundations, philanthropists, and the government.[55] Similarly, they eschewed university affiliation, specifically declining an association with Princeton University in 1974. Unenthusiastic about the possibility of this association, Callahan reminded the board of directors of what he considered the best early advice they had received: "Remain independent. . . . We have not been subject to the policies and priorities of other institutions." He boasted to the board of the virtues of their independence: "Various of our research groups have issued documents and group statements and no doubt will continue to do so in the future. Our institutional independence has . . . been invaluable—we have not had to get the approval of any parent institution for such activities."[56]

Reflecting on the early years, Callahan recalled the "strong impression" that there was "a lot of politics" associated with universities that they needed to avoid:

> We decided earlier on that we had to be a freestanding institution and not be affiliated. . . . We decided we did not want to be burdened with the politics of the university and we would be better off [if, first], we could work with many other institutions, even with universities in a way we could not if we were one of them; and, secondly, that we had this very strong impression that the universities had an enormous amount, a lot of politics that went with them and that this would certainly make life hard, not easy. So we decided to remain independent.[57]

Eventually, the center came to realize, however, that even outside funders were in a position to specify, to some degree, the product for which they were paying. Callahan now acknowledges having underestimated the extent to which even foundations had their own agendas.[58] Even at its earliest consideration, however, some of the limitations associated with foundation funding for specific projects had been anticipated. The goal of the ethics institute would be to find "unrestricted funds" so that it might take its own initiative and venture into "uncharted waters."

> Ideally, it is hoped that, over the years, no more than half of the Institute's income will come from money raised for specific projects; the Institute will need unrestricted funds to insure the continuing possibility of taking its own initiatives and venturing in uncharted waters. The fulfillment of that hope would seem to dictate the need for developing an endowment fund as time goes on.[59]

But locating sources of unrestricted money was a continuous strain. Funding agencies were concerned with finding solutions to specific questions. Problematically for the center, funding agencies, in general, did not consider the desire to examine the underlying values and cultural sources of bioethical questions (a chief center aim) as part of any "solution." It was seen, instead, as unhelpful mind-spinning. In June 1971 Constance Foshay of the U.S. Department of Health, Education, and Welfare explained to the institute that the "counterproductive hand-wringing attitude within the group" did not sit well with public administrators: "Public administrators are in the business of solving problems, not just analyzing them. . . . Unless Institute endeavors can be focused not only on analysis of problems, but also on their resolution, public administrators . . . are not going to be very interested in the Institute."[60]

Much more available than the unrestricted grant were grants for projects that required specific advice on how to employ a technology or administer a procedure ethically. For these types of grants, the basic virtue of the production and use of the technology in some form was always assumed. Hastings Center fellow Harold Green suggested the limitations of this situation at a January 1974 meeting of the fellows. The minutes

summarized Green's belief that "specific funded projects" pre-
cluded an examination of "troublesome underlying problems":

> The Institute has been remarkably successful in obtaining grants
> to finance its research activities and this in turn has been reflect-
> ed in the Institute's dramatic growth. He noted, however, that the
> financing of growth through grants forces the Institute to con-
> centrate its activities on specific funded projects requiring a tan-
> gible product at the conclusion of the project. This has drawbacks
> in that it requires all energies to be devoted to ultimate products
> thereby deterring reflection on troublesome underlying prob-
> lems, creating uncertainty among the Institute staff, and gener-
> ally making long-term commitments difficult.... He reported,
> the board of Directors, which has been considering the problem,
> is embarking on a development or fund-raising program. The ob-
> jective is to relieve the current dependence on product-oriented
> work.[61]

The funding-fueled genesis of the "ethics management" approach was responsible, in part, for the creation of one of the
most emblematic devices generated by the bioethics movement,
the development of "guidelines." A Hastings Center recommendation rarely calls for a moratorium on a procedure or the
discontinuation of a program. Instead, a task force will offer
suggestions for how to proceed without violating moral principles. Callahan's 1992 characterization of how bioethics functions suggests the limitations of this approach: "Tell us what
you want to do, and we'll tell you how to do it ethically."[62] Such
a method, he noted, carried a built-in bias that allowed for a critique of the means of medicine but not the ends of medicine.
Extant memos from a Hastings Center study on population control offer a glimpse of the frustrations involved in working on
projects specifically designated by the funding agency.

In 1973 the United Nations invited the Hastings Center to
submit a proposal to study the politically sensitive area of international population control. This area was full of controversy, especially vivid in the accusations made by Third World
countries. These countries charged that the family planning
policies that developed nations were exporting to underdevel-

oped nations actually constituted a form of genocide. From the beginning, the compromises that were necessary to undertake the study disturbed members of the Hastings Center project. Information concerning the project itself is not always clear from the memos. Nor do the memos yield the specific compromises in question. The unease felt by Hastings Center members over the nature of these compromises, however, is unmistakable. They felt hindered from examining underlying ethical problems—above all, issues questioning the virtue of developed nations exporting any family planning programs to Third World countries at all.

Project manager of the center's program in values and population, Donald Warwick, was the principal coordinator for the international project. The project was to be conducted in Mexico, Indonesia, Kenya, and the Philippines. In a memo to Callahan, Warwick discusses a "sticky" problem—that the Mexican president may have been pressured to accept policies foisted upon his country by the U.S.-dominated World Bank. He also suggests that one method for dealing with the problem would be to omit the "sensitive" part of the proposal altogether:

> It is widely rumored that the [Mexican] President did his turnabout toward an anti-natalist policy in response to pressures from the World Bank. Of course, no self-respecting Mexican president would want to look like the puppet of the US-dominated Bank, so that the official version is that this was a purely "Mexican" decision. Hence the sensitivity of that part of our proposal dealing with international agencies. . . . I think this can be handled without compromising our principles or interests simply by omission. I also suspect . . . that this will be a general problem, particularly given the need for governmental approval of projects of this type with the UN.[63]

A subsequent memo from Warwick also alludes to the problem: "For obvious reasons the President and those around him aren't going to admit, if indeed it's true, that they were bludgeoned into changing their policies by McNamara and the World Bank."[64]

Robert Veatch, disturbed by such considerations, argued for a reassessment of their role in the project.

> The problem of the Mexican president is terribly interesting and relevant to the study and to bracket such considerations because they are too hot will really hurt the value of the project substantially. . . .
>
> . . . The Mexican political situation seems very sticky. We already know that the same is true in Egypt, Kenya, and Indonesia and perhaps elsewhere. I wonder if we could not continue the support of the UNFPA study in a minimal way . . . we could do much more by developing an additional, much more modest project . . . of the moral foundations of international population aid and then preparing some guidelines without the political and disciplinary baggage the UNFPA study is introducing.[65]

Callahan concurred with Veatch, fearing that one of the main focuses for doing the study would be compromised, namely, "the ethics of the politics of international aid." In the spring of 1974, after further negotiations, Warwick shared with Callahan an idea for an article based on some newfound wisdom: "Some day, *after* the contract is approved, we should really write up an article about the politics of setting up this project. . . . it's all instructive about why the most interesting population questions don't get studied."[66] Perhaps persuaded by Warwick's analysis that "not doing the study would itself be a compromise," the institute overlooked its hesitations, undertook the project and devised guidelines.[67]

During the following year, the repercussions of trying to examine underlying issues that could threaten funded programs (as opposed to simply devising guidelines) soon became apparent. Events in 1975 began what Callahan later dubbed "the Ethics Backlash," and underscored how adhering to the ethics-management and guidelines-development style of analysis was crucial for the center's survival. "There is a strong suspicion," Callahan explained,

> among many in the biomedical community that much of the new concern for ethics represents a disguised or latent form of anti-

scientific or antitechnological feeling; that the label of ethical analysis is nothing but a cover for attacks on the personal morality of researchers or physicians; that those in ethics display a consistent ideological bias in favor of the protection of individual rights at the expense of that general welfare which medical research and progress can bring.[68]

The institute planned to discuss repercussions from this backlash at its June 13, 1975, meeting. "We have received indirect information from a number of sources," began an agenda memo, "that the opposition in some scientific and medical quarters to ethical analysis has now become intense enough that it could threaten future support for the field from some major Federal science agencies."[69] Indeed the backlash already had affected the center directly when it lost a major grant from the National Institutes of Health, due in large part to what NIH perceived as ideological differences between it and the center. When the three-year NIH grant expired, the Hastings Center genetics group applied for renewal. The proposal projected studies of the ethical, legal, and public policy implications of genetic counseling, prospective genetic screening of newborns, prenatal diagnosis, and the human uses of molecular genetics (e.g., gene therapy). NIH rejected the proposal. Moreover, "the past, present, and likely future work of the group was . . . roundly criticized and condemned."[70] As part of their rationale for rejection, NIH on-site reviewers characterized the institute in a manner that echoed the concerns of others who believed they detected an antitechnological bias among institute members—that is, they felt that the center demonstrated a preoccupation with individual rights:

> The discussants began to converge toward a common viewpoint, i.e., that the individual's needs take precedence over those of society. The reviewers felt that the Group somewhat favors a monolithic philosophical goal that may be antithetical to the "greatest good for the greatest number." In attempts to prevent all individuals from stigmatization, invasion of privacy, and even anxiety, societal goals were submerged or minimized.[71]

Significantly, NIH faulted the center for not promulgating guidelines to assist screening agencies. To rebut this charge, Callahan argued that in the original proposal for the grant awarded previously they had, indeed, mentioned devising codes. At that time, it was NIH that had been cautious about the development of codes. To deal proactively with what they perceived to be the Hastings Center's antitechnological bias, NIH extracted an agreement from the institute to remain "analytic-descriptive" rather than "prescriptive."[72] By 1975, however, NIH changed its mind; now it required guidelines. Callahan's frank defense against NIH criticisms of articles in the center's book on mass genetic screening alludes to why the NIH changed its position. The reviewer's judgments, according to Callahan, reflected

> an erroneous view of what the genetics group's efforts have all been about, which was to explore alternatives, not hand down directives. The present state of the law suggests that there are better and worse ways to set up screening procedures, but it may simply be true that there are no "essential" or correct procedures. The purpose of the article was to identify some legal considerations that might be relevant, not to arrive at a definitive code of behavior.[73]

Promulgating guidelines carried the threat of compelling conformity to external regulation. But, as Callahan understood, biomedical interests feared the far greater threat of a recommendation to prohibit the technological application under study altogether.[74] In the years after 1975 there were few, if any, recorded debates about technology or whether the center should be an activist organization. A type of ethical analysis that advised how to proceed, not whether to proceed, was in place. In the years following 1975, the organization directed its chief energies toward fund-raising (eventually undertaking an endowment campaign), which promised to give the institute the greatest independence. In 1979 Gaylin emphasized its benign posture: "We are not a Nader-type organization and we are not doctrinaire. We don't see ourselves resolving moral dilemmas. But we want to ensure that when an issue reaches the public marketplace, it won't be greeted with hysteria."[75] His remarks also suggest the way in which the Hastings Center came to function, finally, as a pro-technology agency, fostering quiescence over technology-induced social issues that might otherwise "be greeted with hysteria."

Callahan has not been satisfied with the ethics management approach of bioethics. During the highly critical cultural environment of the late sixties and early seventies, Callahan urged a greater temperance among critics of technology. During the more quiescent 1990s, however, he seeks to rekindle some of the spirit of challenge he believes has been lost. In his 1973 work, *The Tyranny of Survival,* Callahan argues for a "science of limits," a science that "seeks less for what should be done than for those boundaries which should not be transgressed in the process." But he was careful to entreat readers to seek a position more moderate than those offered by the "preachers against technology":

> Broadsides against the abuses of technology, however accurate, have accomplished no notable change nor do they promise to. Lewis Mumford, Theodore Roszak and Jacques Ellul, preachers against technology, make provocative bedtime reading, but little more than that, primarily because they fail to take account of the psychological reality principle of technology—that contemporary man cannot and will not live without technology. Failing to see that, they build their critiques on the illusion that things could be otherwise. Of course they could, but not in our world in our time . . . [76]

More recent essays demonstrate a frustration with the course his suggested moderation has taken.

In a November 1990 draft of an essay, "The Future of the Hastings Center: Reflections and Proposals," Callahan discussed the need for redirection. During the 1980s, he believed that bioethics became something of a "standard discipline, aspiring to method, rigor, and recognizable boundaries." Whether this was a good or bad development, it was clear to Callahan that "the broader, less manageable issues" were pushed aside in the process. He also felt that, "Increased pressure to satisfy peer-oriented academic standards, on the one hand, or to produce usable, 'practical' results on the other, worked in the same direction." The field had acquired an ideological focus with a somewhat narrow scope. "With the emergence of ethics committees, ethics consultants, and a widespread demand for ethicists as participants in policy-setting and rule-making, the field is," according to Callahan, "heavily service-oriented, pragmatic, responsive to clinical and legal needs, and resolutely middle-of-the-road in its ideology." Believing that bioethicists had done a better job on "broader issues" in the 1970s than in the 1980s, Callahan urged a reach that will go beyond the "regulatory ethics" of professionalized bioethics. He coaxed colleagues to ask "large and demanding questions . . . even at the risk of seeming less practical and focused than practitioners and policymakers often want."[77]

Ever mindful of pragmatic and financial necessities, he did not recommend shunning regulatory ethics. While counseling that they remain open to educational, regulatory, and policy-making projects, however, he clearly proposed a focus on the ends of medicine. "We should return to some of our earlier aspirations and refresh them, particularly an effort to relate issues of bioethics to the larger social, political, and moral problems facing the nation, indeed the world." His call, in 1990, was for "a fresh public identity and revivified internal sense of mission and direction." He exhorted the center to assume leadership with this expanded role for bioethics: "We should quite deliberately see ourselves as the leader of the leaders, that group which most specifically tries to work with, and enrich, those who run the other institutions."[78]

In a June 1996 article, "Calling Scientific Ideology to Account," Callahan went farther in his call for a renewed sense of purpose. "Science needs," he urged, "a kind of loyal opposition."

> This kind of opposition need not and should not entail hostility to the scientific method, to the investment of money in scientific research, or to the hope that scientific knowledge can make life

better for us. ... What it does entail is a relentless skepticism toward the view that science is the single and greatest key to human progress, that scientific knowledge is the only valid form of knowledge, and that some combination of science and the market is the way to increased prosperity and well-being for all.

He explained that a loyal opposition "would not let the claims and triumphalism of the scientific establishment go unchallenged. It would treat that establishment with respect, but it would fully understand that it is an *establishment* intent on promoting its own cause and blowing its own horn, critical of its opponents and naysayers, and of course never satisfied with the funds available to it."[79]

Callahan regretted that bioethics had fallen short of functioning as this loyal opposition. "There was a time," he confided, "when I hoped my own field, bioethics, might serve as the loyal opposition to scientific ideology."

> In its early days in the 1960s and 1970s, many of those first drawn to it were alarmed by the apparently unthinking way in which biomedical knowledge and technologies were being taken up and disseminated. It seemed important to examine not only the ethical dilemmas generated by a considerable portion of the scientific advances but also to ask some basic questions about the moral premises of the entire enterprise of unrelenting biomedical progress.

Callahan suggested two reason for this failure. First, "most of those who have come into the field have accepted scientific ideology as much as most scientists." Second, "not many people in bioethics ... care to be thought of as cranks, and there is no faster way to gain that label than to raise questions about the scientific enterprise as a whole." He concluded glumly that "bioethicists, on the whole, have become good team players."[80]

One of the higher profile areas of current bioethical analysis is the Human Genome Project. In a 1996 consideration of the nature of bioethics, Callahan was skeptical about the scope of this ethical examination. "It is hardly likely," he surmised, "that the National Institutes of Health (NIH) Human Genome Project would have set aside 5 percent of its annual budget for the Ethical, Legal, and Social Implications program if there had been even the faintest likelihood it would turn into a source of trouble and opposition; and it indeed hasn't."[81]

By their financial rewards and punishments, funding institutions insured the development of a public discourse of ethics, which deemphasized the political. The ethics method of analysis and expression eventually came to mute the dynamic of power in relations—between patient and physician, subject and researcher, citizen and country, nation and nation. In fostering a shift in public discourse from the political to the ethical, bioethics helped to reduce public opportunities for the expression of outrage and limited possibilities for demanding more dramatic change. With the flourishing of bioethics, banner expressions of public life in the 1960s and early 1970s regarding challenges to cultural authority began to wane. How bioethical interests achieved the elision of the political into the ethical is exemplified by the next chapter's discussion.

Chapter Three

Redefining Death
in America, 1968

Professional medicine's request for advice from other disciplines in the 1950s and 1960s was part of the first glimmer of a
demand for bioethics. Throughout the nineteenth century and
into the twentieth, the medical profession had worked ardu-
ously, sometimes acrimoniously, to build and safeguard its autonomy. What so troubled post–World War II physicians that
they broke with this tradition? The coincidence of two separate
trajectories of modern medicine in the late 1960s helped to assure a continuing demand for bioethics. The first, dating to the
1950s, was the anguish of physicians and patients trapped in the
desolate gulf between the promise of medical cure and the reality of therapeutic limitations: in trying to save lives, doctors
often only prolonged dying. The second trajectory, developing
especially in the late 1960s, was professional medicine's desire
to protect itself from the legal liabilities of conducting medical
research—research sometimes dependent upon prolonging dying. The anguish of the first set of concerns sent doctors seeking advice from theologians. The desires behind the second set
had them scurrying to outpace potential legal barriers that
might threaten continued research. This they did by inoculat-
ing themselves against contagion: seeking advice from legal
sources and newly emerging ethical experts, they could perhaps
preempt surrender of their hard-won professional authority.

During the late 1950s and 1960s members of the public and
members of the profession expressed anxieties about modern
medicine's prolongation of suffering. In January 1957 one woman anonymously shared her distress with readers of the *At-*

lantic Monthly. She wrote poignantly about her husband's death at the hands of institutionalized medical care—a "new way of dying . . . the slow passage via modern medicine." After witnessing the torment of repeated surgery, medications, oxygenation, and intubations, after delirium and unconsciousness, the woman could no longer bear her husband's "torture":

> "They can't do this to you any longer. I must put a stop to it," I cried. . . . When the first doctor came on duty I accosted him and begged that they cease this torture. He explained that except under the most unusual circumstances they had to maintain life while they could. Very well, I thought, if it has to be so, so be it.[1]

But when a nurse arrived with a tray of medication, the author wanted to "kick her tray . . . and knock her from the room." "She was here to snare him back just as he might have reached the other shore. I asked her why. 'Doctor's orders,' she replied. 'I am to give him a hypo.' I staggered out the door; there was nothing else to do." The *Atlantic* introduction indicted "big metropolitan hospitals" for creating "an ordeal" that had "somehow deprived death of its dignity."

That anonymous account, although targeted at the general public, resonated with some physicians who themselves had grown critical of what medicine had wrought. Dr. John Farrell, chairman of surgery at the University of Miami, read the widow's anguished memoir. Regarding the *Atlantic* article, the April *New England Journal of Medicine* editorialized that "this is an article that cannot be summarized. It should be required reading for physicians, and perhaps it is as important that the graduate of 1957 contemplate the issues involved as it is for him to ponder on the oath of Hippocrates itself."[2]

In 1958, at a banquet address to a chapter of the American College of Surgeons, Farrell decided to forgo a "light and amusing" theme befitting the occasion. Instead, he echoed the widow's distress.

> In our pursuit of the scientific aspects of medicine, the art of medicine has sometimes unwittingly and unjustifiably suffered. . . . The death bed scenes I witness are not particularly dignified. The

family is shoved out into the corridor by the physical presence of intravenous stands, suction machines, oxygen tanks and tubes emanating from every natural and several surgically induced orifices. The last words . . . are lost behind an oxygen mask.[3]

Farrell cautioned that he did not have the answer. Instead, he counseled that surgeons acquaint themselves with the "discussions of sociologists and philosophers." Additionally, he urged surgeons to ask themselves "wherein lies the glory of a technical triumph which precipitates economic, social or spiritual bankruptcy?" Farrell's moral qualms echoed periodically in journals; and although the technologies invoked as examples varied, the message was the same: an ever widening list of technologies and procedures was being brought into use inappropriately for the supposed benefit of dying patients.[4]

One of the more dramatic indications of the profession's growing discomfort with this problem, and an illustration of the increasing search for moral guidance beyond its own institutions, occurred in 1957. In that year, the International Congress of Anesthesiology, concerned by ethical problems in the use of resuscitative measures, sought guidance from Pope Pius XII.[5] Troubled anesthesiologists asked for moral instruction to advise physicians as to when they had the right or obligation to begin resuscitative measures on unconscious individuals, and, most important, when they were obligated to cease artificial resuscitative measures.

The pope's response was read at the November congress. A physician must not act without authorization from the patient's family, he counseled, and the family was bound to use ordinary, not "extraordinary" measures to prolong life. According to the Catholic Church's "principle of double effect," one act—specifically, terminating resuscitation—had two effects. The first and desired effect was to end human suffering. The second effect, namely, death, was only the indirect result of a moral act. In such cases, terminating treatment was not only permissible but advisable.[6]

In October 1960 theologian Joseph Fletcher shared with

Harper's readers the experiences of ministers and physicians in dealing with the "heartbreaking struggle over mercy death." Fletcher related several tragic stories of prolonged dying and explained that "the right to die in dignity is a problem raised more often by medicine's successes than by its failures." Imparting a sense of the grotesque nature of the dilemma, he disclosed that it was "an unnerving experience . . . to hear an intern on the terminal ward of a hospital say with defensive gallows humor that he has to 'go water the vegetables' in their beds." Fletcher clarified the distinctions between direct and indirect euthanasia and revealed his preference for "the direct method." "Blind, brute nature imposing an agonized and prolonged death is outrageous to the limit," he urged. "It is the very opposite of a morality that prizes human freedom and loving kindness. . . . Death control, like birth control, is a matter of human dignity."[7]

The next year, a *Medical Tribune* editorial defended the use of "heroic treatment":

> From time to time we are criticized for the overly dramatic and desperate treatment of moribund patients—for surrounding the poor soul with infusions, oxygen, pressor amines, residents, and attendings that the relatives can barely have a glimpse of him amid a forest of equipment. The effort is sourly criticized as a "prolongation of death," not of life, and a plea is made for the dignity of a patient's last hours when he ought to be allowed to die in peace. [However,] heroic treatment can succeed. As a result, quite a few "moribund" patients afterward stride out of the hospital in defiance of any reasonable judgment at the time of admission.[8]

But protests against such defenses were passionate and irrepressible.

Responding to the growing desire to address the dilemma of prolonged dying, in March 1966 physicians held the First National Congress on Medical Ethics and Professionalism in Chicago. Dr. William P. Williamson of the University of Kansas Medical Center admonished physicians that, whether they liked it or not, the doctor's "skills, decisions, and the treatments he renders, often determines life or death for his patient." Dr. Williamson provided a partial list of contemporary measures contributing to the dilemma:

> Improved understanding of body physiology and chemistry, potent drugs, remarkably efficient mechanical respirators, pacemakers, and artificial organs, combined with aggressive medical and nursing care, have saved many lives, cured diseases, and solved many medical problems. Yet, paradoxically, this very progress has created other problems . . . [9]

For Williamson, the problems were not medical but theological, social, and legal and the profession needed nonmedical advice to deal with them:

> Consideration of the moral and spiritual aspects, as well as guidance of the family's thoughts and emotions, are proper functions of the clergy, either rabbi, priest, or minister. Thus the team approach of physicians and clergy working together, with patient and family, is the ideal solution to this problem. At times, other professions may contribute . . . such as lawyer, social worker, or nurse.[10]

Prolonged suffering was becoming a concern at even regional medical meetings. It was the topic, for example, at the Wyandotte County Medical Society, which convened in Kansas City in 1966. Each of the attendees brought a clergyman as guest. The program's panel included a surgeon, a psychiatrist, an internist, a rabbi, a minister, and a priest. The meeting was far from the only one called to address similar moral anxieties. The Department of Medicine and Religion of the American Medical Association (AMA) helped form committees on medicine and religion in each state medical society. In turn, those committees arranged seminars and discussion groups to address the dilemma.[11]

Those efforts were part of concerns growing in intensity since the fifties. The dilemma was not the result of a specific technology or procedure. It was a predicament rooted in a medical cast of mind that measured success by medicine's ability to stall death, even in the face of death's inevitability.

In the late sixties, anxieties over prolonging dying clashed directly with the desires of medical researchers wishing to move ahead with experimentation in the controversial field of organ transplantation. Human experiments in kidney transplantation in the mid-1960s had generated harsh criticism from within the profession.[12] Heart transplantation compounded this controversy with a unique difficulty. Whereas a healthy individual could donate one kidney and still function normally with the remaining kidney, a healthy human being could not donate a heart. Cadaver heart donations had been tried and found unsatisfactory. The best sources for donations were severely brain-damaged individuals attached to respirators. The problem for transplant surgeons, however, was that these individuals were, technically, still alive.

Dr. John H. Kennedy put it bluntly: the donors of early human heart transplants "would not have been pronounced dead [at the time they had been pronounced dead] other than for their role as cardiac donors." In a 1968 letter to the *Journal of the American Medical Association* (*JAMA*), Dr. Kennedy urged the profession to "assume a leadership role" in considering new criteria for death: "If the criteria of clinical death are indeed to become more liberal to meet the demands of a new method of treatment of incurable human disease, the medical profession must assume a leadership role in the careful consideration of these criteria."[13] *JAMA*'s editors concurred that transplantation presented this problem. Unless the criteria for being "finally and irretrievably dead" are changed, "murder has been done."

> Among the many problems surrounding the transplantation of vital organs . . . one of the most critical is this: When is the donor finally and irretrievably dead? When does one dare to relieve him . . . of his heart or his liver . . . ?
>
>   . . . If such organs are taken long after death, their chance of survival in another person is minimized. . . . if they are removed before death can be said to have occurred by the strictest criteria that one can employ, murder has been done.

Two months later, the editor queried: "Can the transplanter afford to wait for a dying organ just to be certain that he is not also a surgical criminal?"[14]

In August 1968 heart transplantation researchers won crucial support: the Ad Hoc Committee of the Harvard Medical School to Examine the Definition of Death codified and publicized in the United States new criteria for determining death—criteria that allowed for removal of a beating human heart.[15] Often referred to as a "redefinition of death," the criteria actually established an additional way of determining death rather than creating a new definition that superseded the old definition in all cases. Cessation of respiration and heartbeat was no longer the sole criterion for determining death. Four new standards were recommended: (1) unreceptivity and unresponsitivity, (2) no movements or breathing, (3) no reflexes, and (4) a flat electroencephalogram. The committee had offered new criteria for declaring someone "brain dead."[16]

The thirteen members of the committee drew on the gravity and tradition of medicine's anxieties about prolonging dying when they explained the motivation for "redefining" death. Justifying the necessity of an unprecedented new definition, the committee's ten doctors, one lawyer, one historian, and one theologian underscored advances in technology as the principal motivation; improvements in resuscitative and supportive measures had led to situations in which individuals with beating hearts could be maintained "alive" even though they were irreversibly brain-damaged, a situation that adversely affected the patient and emotionally and financially drained families and hospitals.

The committee's mention of the problem of resuscitative measures sought to place its motivations along the trajectory of professional concern for human suffering, which had begun more than a decade earlier as a dissenting viewpoint. In so doing, members of the committee obscured the impetus that had, in fact, generated the urgency for redefinition, an impe-

tus mentioned by the committee as a secondary factor—
namely, the concern that obsolete criteria for the definition of
death had led to controversy in obtaining organs for transplantation. This impetus was part and parcel of a separate and,
in this instance, contradictory trajectory: the demands of scientific research. The committee's new criteria sought to limit
controversy surrounding organ procurement by making it legal to remove a beating heart from someone who, by contemporary popular and legal understanding, was still living.[17]

The committee offered two reasons for redefining death,
reasons based on mutually contradictory values. First, members said that a new definition would help reduce the emotionally and financially draining effects of resuscitative measures. That motivation was predicated on a desire to relieve
human suffering for patients and their families. The second
reason was to reduce controversy surrounding organ transplantation. That impetus, however, sought to further scientific
research even at the cost of prolonging human suffering. This
was true for heart transplantation in two ways. First, once the
criteria for irreversible coma went into effect, potential heart
donors could be placed on respirators and left there for *longer*
periods of time in order to help ensure the integrity of the organs. Thus organ procurement could directly contradict the
first reason for redefining death, that is, to reduce suffering.[18]

The second way in which heart transplantation research
prolonged suffering is made clear by its track record during the
late sixties and early seventies. The moral uncertainties surrounding heart transplantation in the 1960s were many. Animal heart transplants had been unsuccessful.[19] Despite the
lack of success with animals, researchers turned to humans.
The early months of this "therapy" had a complete and completely predictable failure rate.[20] Human heart transplantation was, in fact, experimentation on human subjects conducted in complete view of an awestruck public that believed
it was seeing the latest medical miracle.[21] The media sensa-
tionalized the drama and excitement of each procedure, in-
spiring public astonishment and approval.[22] Notices of the

deaths of transplant patients days or weeks later, often described in less dramatic reports, did not dampen popular amazement. The ethics of continuing to offer the procedure was, at best, questionable.

In March 1968, scarcely two months after the world's first human-to-human heart transplant in South Africa, three top United States cardiac specialists called for a moratorium. Dr. George Burch of Tulane University in New Orleans was particularly direct in his denunciation. Dr. Burch would not select any patient for a cardiac transplant because "once you take his own heart out, you know he's going to die.... His new heart will be rejected by his body because we are still unable to suppress the immune reaction."[23]

The Harvard committee's attention to its contradictory rationales eventually helped to shift focus away from the controversy surrounding heart transplantation and toward the role of a reified technology in prolonging suffering—specifically, the role of "improvements in resuscitative and support measures." The medical profession, however, had been dealing with the disturbing effects of artificial life-support systems for well over a decade. Indeed, the pope had offered his solution to the dilemma in 1957, a solution that the Harvard committee had invoked favorably, never explaining why the papal resolution was, for some reason, now insufficient. Why, then, was the problem being readdressed in 1968?

The committee statement gave the impression that advances in respiratory technology were the principal development responsible for generating ethical concern.[24] "From ancient times down to the recent past," the committee instructed, "when the respiration and heart stopped, the brain would die in a few minutes.... This is no longer valid when modern resuscitative and supportive measures are used. These improved activities can now restore life."[25] Chief among the "improved activities" was the mechanical ventilator.

Mechanical ventilation was not, however, a recent development. The iron lung was the technological predecessor to the modern ventilator.[26] Developed in 1929, the apparatus was itself an adaptation of devices developed as early as the late nineteenth century. The patient reclined inside a round steel tank. The head remained outside the chamber while the neck was supported by a tight-fitting rubber collar designed to avoid pressure to the windpipe and voice box. Inside the tank, adjustable pumps applied intermittent positive and negative pressure. The iron lung saved the lives of hundreds of victims of poliomyelitis, whose breathing had been impaired by the paralysis of intercostal muscles and diaphragm. The device was crucial in helping patients through weeks or months of temporary paralysis. But, due to complications and the need for full time nursing, it was an imperfect solution for long-term use.

One of the main innovations leading to the modern ventilator came in 1952 in Denmark when physicians were faced with a catastrophic polio epidemic. After twenty-seven of the first thirty-one patients on respirators died at the Blegdam Hospital, Drs. H. C. A. Lassen and Bjorn Ibsen attempted an innovation: a tracheotomy followed by mechanical ventilation from a manually compressed anesthesia bag that had been adapted for the purpose. In all, 250 medical students worked in daily relays to manually ventilate approximately two hundred patients. Three years later, about twenty-five patients were still being kept alive through ventilation, a technological process that had since become automated.

With modifications the use of mechanical ventilation grew and by the late 1950s the respirator was used to treat illnesses and injuries beyond polio. Statistics from Massachusetts General Hospital are instructive. In 1958, sixty-six patients were kept on ventilators for twenty-four hours or longer. By 1964 the number had grown to four hundred, by 1968 to about nine hundred and by 1982 to approximately two thousand.

The risks of numerous surgical procedures and illnesses were vastly reduced by use of the ventilator. Artificial respiration, clearly, had been one of the major achievements of modern medicine. Yet successes in medicine are rarely, if ever, unqualified. Ventilators, along with renal dialysis, intravenous feeding, and a panoply of other technologies and procedures, did not always assist in restoring health. Often, they either prolonged dying or prolonged life indefinitely in a "vegetative" state.

If difficulty in determining when to terminate resuscitation was a problem as early as the 1950s, why did it become such an acute issue in 1968. Why was redefining death chosen as the solution? Publicly codifying a new definition of death was not a simple solution to a dilemma that had plagued the profession for decades and had, somehow, been overlooked until 1968. As was widely understood within the heart transplantation specialty and as the popular press occasionally reported, the need to redefine death stemmed from the desire to reduce controversy in obtaining organs.

Transplant surgeon Thomas E. Starzl recalled in his memoirs the first time that he heard of the concept of brain death. At the 1966 Ciba Foundation symposium on the ethics of transplantation in London, Guy Alexandre of Louvain, Belgium, told of extracting kidneys from "heart beating cadavers." Starzl was "appalled" upon first hearing about the removal of organs from brain-damaged people with beating hearts. He "envisioned that the care of a trauma victim could be jeopardized by virtue of his or her candidacy to become an organ donor." Ultimately, however, he came to believe that his initial fears were unfounded. He was persuaded that once it became licit to remove organs from brain-dead individuals, it became *more* likely that accident victims (i.e., potential donors) would be placed on respirators than before, thus *increasing* their chances of revival. Placing patients on respirators helped to preserve their organs, and, as Dr. Starzl phrased it: "With the wide acceptance of brain death in the Western world, all injured patients who come to the hospital in a helpless condition could have a fair trial at resuscitation." Given the parallel though inconsistent concern over prolongation of dying, the source of Dr. Starzl's sense of comfort bears a striking irony. He adds the following statement, remarkable in the way it reveals how the demands of scientific research can alter standards of care: "Then, in an orderly way, it can be determined whether these people already were dead but with functioning hearts and lungs, or if they had a chance of restoration of brain function. The quality of care and the discriminate application of such care to terribly damaged people was one of the great fringe benefits of transplantation."[27]

Dr. Starzl described some of the difficulties for surgeons wishing to transplant organs before 1968—difficulties that prompted them to consider redefining death. Before 1968, organs could not be removed until after the ventilator had been disconnected; that left little time to remove organs before they were damaged from oxygen starvation and the "gradually failing ... circulation."

> Rather than trying to maintain a strong heartbeat and good circulation in the cadaver donors, the legal requirement before the end of 1968 was just the opposite. Because all such donors were incapable of breathing if the brain actually had been destroyed, they were supported by ventilators. The steps to donation began with the disconnection of the ventilator, which the public called "pulling the plug." During the five to ten minutes before the heart stopped and death was pronounced, the organs to be transplanted were variably damaged by oxygen starvation and the gradually failing and ultimately absent circulation.
>
> The strategies that could be used to minimize the organ injury under these circumstances were limited.[28]

If "heart beating cadavers" facilitated organ transplantation generally, they were the sine qua non of heart transplantation specifically. Dr. James Appel, president of the American Medical Association from 1965 to 1966, explained the impetus for redefining death when he asked two questions rhetorically. "Should the physician attempt to restore heart action in the donor patient and then after a period of time turn off the resuscitator?" he queried. "If the heart does not continue to beat independently of the machine, should he declare the patient dead and then start the machine again so that there will be some circulation in the organ to be transplanted? These questions have caused the medical scientist and the medical practitioner to wonder if we should not have new criteria to determine legal death."[29] These concerns were widely understood within transplantation research circles during the 1960s. One physician even referred to the brain death criteria as the "new definition of heart donor eligibility."[30]

By the 1970s, the historic motivation behind having sought the brain death criteria was understood by medical researchers, as the 1977 *JAMA* article "An Appraisal of the Criteria of Cerebral Death" makes clear:

> The need for viable organs led surgeons to seek patients with intracranial pathological findings such as head injuries that resulted in a dead brain. However, if the pronouncement of death were delayed until the heart stopped beating, the organs underwent so much deterioration that a successful transplant was jeopardized. Hence, a definition of human death that considered the lack of cerebral function as important as the cessation of cardiac activity was recognized.[31]

The Harvard committee had promulgated its criteria to avoid the kind of controversy that surrounded Dr. Dentin Cooley, a transplant surgeon at Houston's St. Luke's Episcopal Hospital in May 1968. Cooley's team was prepared to remove the heart of thirty-six-year old Clarence Nicks, a welder who had sustained severe brain damage during a brawl the previous month. Lawyers disagreed over whether Nicks's assailant could be charged with homicide if doctors removed the respirator while his heart was still functioning.[32] Would Nicks have been killed by the barroom brawler or by the physicians who removed his heart? Moreover, since Nicks's death would constitute a homicide when he did die, an autopsy would be necessary. It was unclear to Harris County Medical Examiner Joseph Jachimczyk whether he would be able to fulfill the legal requirements of an autopsy if Nicks's heart were removed.[33] He was reluctant, as such, to let the transplant team have the heart. Jachimczyk learned, however, that Nicks's brain no longer functioned and that Nicks had been pronounced dead at 10:30 A.M. on May 7; the "patient" was being kept on a respirator for the purpose of keeping his heart viable for transplantation donation. This being the case, the medical examiner promised that no charges would be brought for interfering with an autopsy were Cooley to go ahead with the transplant, which Cooley did complete.

Later, Dr. Norman Shumway of Stanford University also came up against the law. The family of a homicide victim wanted the victim's heart to be used for transplantation. Stanford Hospital, however, had agreed with the coroner of Santa Clara County that organs would not be taken from homicide victims. Despite this understanding, Stanford Hospital acquired the heart, and Shumway proceeded with the surgery. Lawyers for the defense argued that the defendant did not kill the victim because the victim did not die until his heart was removed. The jury agreed.[34]

In contrast to 1957, when ethical uneasiness prompted a search for guidance from a religious source, Harvard's brain death committee had legal concerns in mind. The preoccupation with avoiding legal conflict is underscored in that subsection of the brain death criteria discussion entitled "Legal Commentary." Here, members cautioned surgeons that patients, designated as donors, should be declared dead before being removed from artificial ventilation because it would provide surgeons with greater "legal protection."

> We recommend the patient be declared dead before any effort is made to take him off a respirator. . . . This declaration should not be delayed until he has been taken off the respirator. . . . The reason for this . . . is that . . . it will provide a greater degree of legal protection to those involved. Otherwise, the physicians would be turning off the respirator on a person who is, under the present strict, technical application of the law, still alive.[35]

Later, the committee suggested how to avoid the appearance of a conflict of interest when declaring death:

> It is further suggested that the decision to declare the person dead, and then to turn off the respirator, be made by physicians not involved in any later effort to transplant organs or tissue from the

deceased individual. This is advisable in order to avoid any appearance of self-interest by the physicians involved.[36]

Because medical consensus determines legal standards of care and definitions, the acceptance of the Harvard ad hoc committee's redefinition of death by medical practitioners generally was an important tool in stemming the tide of legal liabilities. Also important were state statutes that endorsed the brain death criteria, although the Harvard committee had hoped to stave off the necessity of legislative action by encouraging the medical profession generally to adopt the brain death criteria.[37] Owing to the likes of legal nettles that hampered Doctors Cooley and Shumway, however, legislation did prove necessary. The Kansas legislature passed the first such statute in 1971.[38] Dr. Shumway's experience was instrumental in California's adoption of a similar statute in 1973.

The month following publication of the Harvard criteria, the Second International Congress of the Transplantation Society met in New York. Dr. F. C. Spencer of New York University Medical Center chaired a symposium on the use of donor hearts from "brain dead" individuals. Without reference to the Harvard committee, he was, essentially, in agreement with it. A heart could be kept beating, he noted, for hours or days after "death" using artificial means. Constraining transplant surgeons to wait until the donor heart ceased to contract before removing it was pointless, he claimed. Such hesitation compromised the viability of the organ even while the organ was no longer of any use to the donor. A number of surgeons, including Dr. Dentin Cooley, who by then had completed eleven heart transplants, agreed.[39]

The Harvard committee's published statement begged a question that neither it nor the Second International Congress of the Transplantation Society addressed: had physicians who discontinued resuscitative measures before 1968 killed their patients? According to the pope, they had not. According to medical customary practice, which is the foundation for legal standards of care, they had not. Indeed, had the possibility of this act being considered killing been the only cause for concern, there would have been little legal reason for the profession to codify criteria for determining death at all. What had been customary medical practice for well over a decade could remain so, and with it legal standards could remain. As a 1968 *JAMA* article explained to its readers, "doctors are in a position to fashion their own law to deal with cases of prolongation of life. By establishing customary standards, they may determine the expectations of their patients and thus regulate the understanding and the relationship between doctor and patient. And by regulating that relationship, they may control their legal obligations."[40] From the point of view of medical research interests, what necessitated action was the possible conflict of interest caused by heart transplantation and fears over the criminalization of efforts to procure organs.

More than a medical response to a technologically induced moral problem, "brain death" was an artifice of legal self-protection. It was designed to protect professional medicine against the possibility that the public would perceive a potential conflict of interest and become alarmed—a conflict between the physician's responsibility to care for the sick and dying and the demands of medical research to procure organs for transplant.

Redefining death was not simply a matter of technical or professional medical concern, however. When the Harvard committee promulgated the brain death criteria in the pages of *JAMA*, it simultaneously published a summary of "A Definition of Irreversible Coma" in the *New York Times*. Public acceptance of irreversible coma as "brain death" was crucial. If the legal ramifications surrounding brain death could be addressed by urging new customary practice, it was not so clear to proponents of the new criteria how public anxieties could be managed. As one Washington public health official remarked, "I have a horrible vision of ghouls hovering over an accident victim with long knives unsheathed, waiting to take out his organs as soon as he is pronounced dead."[41] The medical research community in the United States was mindful of the experience in other countries where transplantation and the redefinition of death had been debated. In May 1966 Dr. Clarence Crafoord had provoked public outcry in Sweden over the "cannibalizing" of human beings for spare parts when he suggested that persons be declared dead when flat electroencephalograms indicated irrevocable brain damage.[42] The potential for public protest was a serious concern for champions of transplantation research.

Several months after the Harvard committee announced its criteria for irreversible coma or "brain death," an article in *JAMA* noted that cardiac transplants had given the diagnosis of death "a new public dimension." The authors urged that systematic attempts be made "to assess public concern and to involve the public in a dialogue about the vital issues raised by new concepts dealing with the diagnosis of death." They stressed that the public was, indeed, concerned about how death was determined. "The information emanating from the mass media regarding heart transplants had stimulated much of this thought," they suggested further. "The movement toward upgrading the diagnosis of death," they concluded, "will need to be preceded by some program of public education."[43]

The authors analyzed literature on eighteenth- and nineteenth-century fears regarding diagnosing death and the problem of premature burial. Advising that it was likely that the public would, once again, become concerned and involved, they urged an effort to mold public opinion. Without professional guidance, public interest could become problematic.

> This involvement may be misguided or even ludicrous, but it can become forceful and even restrictive. . . .
>
> Just as styles of dissent change from period to period, so do styles of communication and the methods of molding public opinion. The form and style of the public dialogue on medical ethics had not completely taken shape, but it would appear that the dialogue will need to be candid as well as broadly based. . . .
>
> A public dialogue can and should become an important instrument in developing a climate within which medical progress and community welfare can be maximized.[44]

Organized medicine had already begun the dialogue. The AMA's popular journal, *Today's Health*, spoke to public fears over transplantation in an April 1968 article by the chief *JAMA* science writer. Heart transplantation, Ulys Yates explained, was unlikely to bring relief to any but a few patients, partly because of the nature of heart disease but also because of the shortage of transplantable hearts. The author reassured readers that donor hearts would not be removed prematurely: "Despite worry expressed in some quarters that the need for donor hearts might lead surgeons to remove the organ from a potential donor before he is actually dead, this is not likely to happen," he explained. "Irreversible brain damage occurs within five minutes after the heart stops beating. And any surgeon would wait at least this long before starting to remove the donor heart."[45]

Efforts on the part of medical research interests could not fully shape public or medical professional opinion on transplantation and the redefinition of death, however. The New York State Legislature, for example, appointed a nine-member commission to hold public hearings to determine a "precise definition" of death for organ transplant donors. New York Supreme Court Justice J. Irwin Shaw, chairman of the Temporary Commission on Vital Organ Transplant, said the hearings were the result of "a flurry of heart transplants."[46] By March 1969 *Science Digest* reported that "the heart transplant era is provoking fears among some people that they might be buried alive. . . . They are reading about sophisticated new ways of determining death . . . [and] are becoming worried over premature burial."[47]

Published discord within the medical community was modest, but it is clear that at least some physicians were unconvinced of either a legitimate need, or the possession of sufficient knowledge for establishing brain death as a definition of death. One way to avoid public uneasiness was simply not to make transplantation requirements a priority at all; death should be determined as death had been determined before harvesting organs became a priority. This was the advice of Dr. J. Russell Elkinton, editor of the *Annals of Internal Medicine*, in March 1968. Elkinton worried that the potential conflict of interest of a physician treating a near-death patient whose organs were wanted for transplantation might lead to jeopardy for that patient:

> Because of the need at present to do the transplantation as quickly as possible after the death of the donor, his care as a patient may be jeopardized or the moment of death prematurely anticipated. The first of these hazards might be diminished by keeping the primary responsibility for his care in the hands of doctors other than the transplantation team. . . . Avoiding the second hazard turns . . . on careful definition of the time of death.

Elkinton explained that conventional requirements for determining neurological signs of death included monitoring a flat electroencephalograph for a period of time that was inconsistent with ensuring that organs would be viable for transplantation. Nevertheless, writing five months before the publication of the Harvard brain death criteria, Dr. Elkinton recommended that "the more conventional criteria of cessation of heartbeat and respiration must be used by the physician responsible for such dying patients (potential donors). We do not want to apply a double ethical standard: one for the unconscious patient with a head injury who is not being considered as a possible donor of an organ and another for the same kind of patient who is."

If there was any chance at all that the patient might recover after resuscitation, Elkinton argued, the patient must be maintained on the respirator for the number of days "necessary to establish that the minimum neurological criteria had been met for irreversible damage and death of the central nervous system." This must be done even though such precautions "may be less than satisfactory to the doctors responsible for the recipient patient"—in other words, even though the transplant doctors would not be able to salvage viable organs. While his recommendation was an unfortunate one for the doctors of recipient transplant patients, Elkinton believed that the public at large would find his approach reassuring, "and the public will need reassurance as more dying patients are sought as a source of organs for transplantation."[48]

Dr. W. N. Hubbard, dean of the University of Michigan Medical School in 1968, was dissatisfied with the idea of both mechanical respiration and organ transplantation. He echoed professional anxieties about prolonging suffering when he spoke of organ transplantation and life-support systems as ways modern medicine had of protracting dying. In his June address to the graduating class of the Albany Medical College of Union University, Hubbard stressed that sometimes a patient's right to die outweighed the physician's "right" to attempt to extend the patient's life through surgeries like organ transplants. And he implicated the inappropriate use of life-support systems as another example of medical imprudence. "The physician must beware of treating his own anxiety that death represents his personal failure by unrestrained use of life-support systems."[49]

The director of the National Heart Institute in Bethesda, Maryland, Dr. Theodore Cooper, called for additional research on determining death. Dr. James Hardy from the University of Mississippi dissented from even considering new criteria on the grounds that many communities were not likely to accept readily the removal of a beating heart.[50] The lurid sarcasm of Dr. Edward Shaw from the University of California's Medical Center in San Francisco suggests that although dissent may have been infrequent, it was lively. The perfect solution to the donor problem, according to Dr. Shaw, would be to employ the French guillotine. Such a device would provide a donor "in whom the precise instant at which cerebrum function ceases is beyond doubt." Moreover, there would be no "possibility of restoration." Shaw averred that such a solution was "wholly revolting" but concluded that it was not more so than some of the suggestions he had read.[51]

There was some dissent on an international level as well. In August 1968 at the World Medical Assembly in Sydney, Australia, members from sixty countries discussed the question of organ donation and the definition of death. Britain's Dr. Martin Ware related how in four cases in Edinburgh, an electroencephalograph indicated lack of brain activity, yet all four patients recovered. Netherlands physician G. Dekker reported a similar case in which the machine registered no brain activity for four days, after which the patient recovered. Harking back to historic difficulties in determining death, Sir Leonard Mallen of Australia, the newly elected president of the assembly, commented that not even rigor mortis was a reliable indication of death in the light of "new" resuscitative methods.[52]

Although medical research interests were unable fully to contain criticism, efforts to negotiate through hostile opinion and avoid debilitating protest were not ineffective. For its part, the public remained confused about but not monolithically opposed to medical science with respect to organ transplantation and the redefinition of death. Even today, how death should be defined remains an unsettled matter of discussion and subdued debate (see the Epilogue). At the historic moment of 1968, one year in an incendiary decade when challenges to professional authority were both common and shrill, even a limited embrace of controversial transplantation experimentation spelled a type of success for the backers of this research who promulgated the brain death criteria.

When the editor of the *New York Times* penned his opinion on brain death, he either ignored or failed to see that in providing a solution to the problem of organ procural for heart transplants, the brain death criteria had been a response to the controversial needs of experimentation on human subjects. Noting that the Harvard committee also had organ procurement in mind, the editor interpreted its efforts chiefly as an effort to help free "the human vegetables" among us:

> As old as medicine is the question of what to do about the human vegetable, the individual who—because of brain injury or disease—goes into irreversible coma while his heartbeat and metabolism continue with external aid. Sometimes these living corpses have "survived" for years, draining the financial and emotional resources of their families.[53]

The editor exonerated the criteria because, as he saw it, "Adoption of this proposal would authorize physicians in . . .

tragic cases to halt the artificial respiration or other means being employed to continue 'life.'" Having accepted the committee's rendition of motivations, he speculated that physicians must have been terminating treatment quietly in such
cases in the past—"but always at the risk of being accused of
murder." "The redefinition now suggested," he supposed,
"would end that problem." The editor could not have been
more wrong on at least one score: before 1968, no physician had
been accused of murder for terminating resuscitative measures.[54]

Not all readers of the *New York Times* were persuaded
by its editor's imprimatur of the new death criteria. For
S. Drucker, "The demand of the Harvard medical panel for a
new definition of death, so that removal of organs from comatose patients can be legalized, indicates the extent to which
society has abdicated to surgical experiments. . . . Science is altering, transferring, and mutilating life as if medical experimentation were explicitly sanctioned by the terms of medical
license. . . . No dying person can give his consent to be mur-
dered by a surgeon, and no physician can be safely trusted to
pronounce a live person dead on the basis of semantic juggling
of words."[55] But Fred Anderson, writing for the *New Republic,* chose neither to condemn nor to approve (additionally, his
reaction demonstrates the readiness that many felt for an interdisciplinary expertise to grasp such ethical nettles—something bioethics was soon to promise): "The opinion of the Harvard committee on brain death is not clearly right or wrong;
the question is still open, and before it can be answered adequately the best thoughts of many, including theologians,
philosophers, economists and jurists, along with physicians
must be heard."[56]

Efforts to negotiate smooth public acceptance of brain death
criteria ran into continuing difficulties. The Harvard committee's recommendation that brain death be accepted uniformly
by the profession as customary medical practice was not an authoritative legal act. The *legal* status of the committee's suggestions had yet to be determined. The proclamation of brain

death criteria could not, on its own, prevent litigation similar to the Nicks case from plaguing transplant surgeons. Whether a given state court should accept the brain death criteria continued to play a role in litigation into the 1970s.[57]

Another continuing difficulty for transplant research interests involved stories of accident victims reported to have "come alive" after being declared dead, sometimes just as surgeons were attempting to remove their organs. In May 1974 a sixty-four-year-old man from Birmingham, England, had been presumed dead for an hour after sustaining serious brain injury from a car accident. Dr. Anthony Barnes and his assistant, Dr. Susan Padmore, had already opened the victim's abdomen when the man twitched and coughed. "Did he do that, or did you?" Dr. Barnes asked his assistant. Dr. Padmore pointed to the "donor" and replied, "No, he did that!" They immediately closed the incision.[58] Similarly, after suffering head injuries on July 9, 1974, twelve-year-old Jolene Kennedy of North Carolina was being prepared to have her eyes and kidneys removed for transplantation. Before physicians could remove the organs, however, Jolene's hand moved and she began to breathe.[59]

The public remained conflicted about medical science with respect to organ transplantation. If medicine generated fear, it also offered hope. Its limitations may have been apparent but so too were its promises. There was no ground swell of popular opinion critical of heart transplantation as questionable experimentation on human subjects; no public cries for a moratorium ensued. Braking efforts had come from within the profession itself.[60]

Similarly, despite early indications of an impending controversy, there was no persistent public outrage over the fact that the urgency of redefining death had derived not from concern over death with dignity but from the inconsistent desire for transplantable organs. Perhaps a more significant indicator of popular mistrust—a "silent protest"—lay in the fact that, despite pleas for organ donations and promulgation of such facilitative legal devices as the Uniform Anatomical Gift Act, organ donation did not become a popular phenomenon. A British government report noted that after dramatic publicity over transplant cases, the supply of donor kidneys would drop sharply. Similar drops were noted in the United States. In fact, the *Hastings Center Report* noted that heart transplantations had dwindled—in part because the supply had "dried up."[61]

As backers of transplantation research promulgated criteria for irreversible coma and the public considered them, bioethicists analyzed the motivations behind and implications of this new brain death definition. Joseph Fletcher, theologian and pioneering medical ethicist, presented the case for medical progress in organ transplantation in its most extreme version. Fletcher, commenting before the Harvard committee published its criteria for brain death, saw no point in waiting for individuals to "die" before designating them as potential organ donors. When death was "positively" inevitable, Fletcher saw little problem with the idea of speeding up the donor's death if to do so meant providing valuable life to another human being.[62] Additionally, Fletcher would not let lack of familial consent be a barrier to obtaining needed organs. "If there is no way to get permission from heirs in time," Fletcher advised, "I think properly qualified medical people and public officials should proceed with transplants that have a good chance of preserving or substantially extending life."[63]

Hans Jonas and Paul Ramsey, two of Fletcher's colleagues and members of the fledgling Institute of Society, Ethics and the Life Sciences (the Hastings Center), saw matters differently. Jonas found strong reasons for dissatisfaction with the Harvard criteria for irreversible coma (brain death). The motivation behind the Harvard criteria was, according to Jonas, to seek "permission not to turn off the respirator, but, on the contrary, to keep it on and thereby maintain the body in a state of what would have been 'life' by the older definition . . . so as to get at his organs and tissues under the ideal conditions of what would previously have been 'vivisection.'" Jonas rejected the Harvard criteria completely:

> *Since we do not know the exact borderline between life and death,*
> nothing less than the maximum definition of death will do—

brain death plus heart death plus any other indication that may be pertinent—before final violence is allowed to be done. . . .

When only permanent coma can be gained with the artificial sustaining of functions, by all means turn off the respirator, the stimulator, any sustaining artifice, and let the patient die; but let him die all the way. Do not, instead, arrest the process and start using him as a mine while, with your own help and cunning he is still kept this side of what may in truth be the final line. Who is to say that a shock, a final trauma, is not administered to a sensitivity diffusely situated elsewhere than in the brain and still vulnerable to human suffering? a sensitivity that we ourselves have been keeping alive? No fiat of definition can settle the question. But I wish to emphasize that the question of possible suffering (easily brushed aside by a sufficient show of reassuring expert consensus) is merely a subsidiary and not the real point of my argument; this, to reiterate, turns on the indeterminacy of the boundaries between *life and death*, not between sensitivity and insensitivity, and bids us to lean toward a maximal rather than a minimal determination of death in an area of basic uncertainty.[64]

Theologian and bioethicist Paul Ramsey did not go as far as Jonas in rejecting the criteria completely but did believe that the "highest loyalty" was owed to the primary patient (as opposed to a potential organ recipient). This being the case, "neither of the procedures for stating death nor a decision that death has occurred [should be] distorted by any reference to someone else's need for organs. This loyalty to the primary patient may call for waiting longer or stopping earlier than may be the case if one patient's dying . . . is made adjunct to another's life."[65]

Jonas and Ramsey were members of the Hastings Center's death and dying task force. Also a member of the group was Dr. Henry K. Beecher, an anesthesiologist. Beecher had been the driving force behind organizing the Harvard committee that fostered the brain death criteria. Himself a "bioethicist," Beecher had helped define the emergent field. In his 1966 *New England Journal of Medicine* article, "Ethics and Clinical Research," Beecher helped launch the furor over human experimentation by "blowing the whistle" on unethically designed research projects, which medical journals, nevertheless, had published. On the issue of brain death, however, he found himself in disagreement with a number of his Hastings Center colleagues on the death and dying task force. Beecher acknowledged this disagreement and found it collegial: "I thrive on controversy and I take no exception to the spirit in which various criticisms were made. I suppose one difficulty is that Paul Ramsey and I approach each other from different disciplines, hence different conclusions are arrived at. But, I agree with you, that is the nice thing about our group. We can debate these matters."[66]

In April 1970 Hastings Center founder Daniel Callahan functioned as mediator between Beecher and his critics at the center. Callahan asked Beecher to "moderate the somewhat polemical tone" of certain sections of a draft of a paper concerning the redefinition of death. Himself a professed newcomer to the debate, Callahan urged Beecher to reassure members of the task force regarding fears about "organ snatchers":

> I gather that the main worry is that the freedom gained for some of the parties (the potential organ recipient and those making the judgment that death has occurred) will be at the expense of the moribund person.... Perhaps ... , you could sketch out something in the way of a recommended policy, one that could serve to set at rest the worries of those fearful of "organ snatchers."

Callahan underscored the belief that it was the "transplant possibilities" that had "pressed the matter" on them:

> On the crucial point of relating the new definition to the transplant problem—which Ramsey and some others would like to see kept totally separate—I think Morrison at one meeting provided a good insight: it has been the transplant possibilities which has pressed the matter upon us. That means that it becomes very difficult to keep the two issues separate, even if one would, ideally, like to do so.[67]

Beecher's response was emphatic: the idea that the new definition of death "was created by the need for organs" was an "unfounded conclusion":

The new definition was formulated because, first, the old definition of death is quite illogical and inadequate (i.e. heart death);
secondly, with improvement in resuscitative techniques we have
an increased and increasing number of decerebrated individuals
on the hospital wards. For example, respiratory units are being established in all major hospitals. We simply have to come to some
new decisions as to how long we can continue to expend our resources to no avail. The new definition of death is simply a case
of thinking through and establishing orderly procedure for dealing with these problems. Third, there is a benefit to the needy so
that their lives may be greatly extended by the use of organs no
longer of value to the original proprietor.[68]

The account that Beecher offers here gives no indication of
whether there was, within the profession or without, concern
over problems associated with mechanical ventilation significant enough to warrant redefining death. It is not implausible
that for Beecher, an anesthesiologist (and therefore familiar
with resuscitative difficulties) and an ethicist, the *anticipation*
of ethical unrest that might accompany increasing use of resuscitative techniques was sufficient reason to undertake closer
scrutiny of the matter. Such a singular anticipation did not,
however, reflect widespread concern. Moreover, Beecher's account does not explain why publicly redefining death would
have been seen as the preferred method for dealing with the
problem of severely brain-damaged people on artificial respiration—especially since the brain death criteria, technically
speaking, seemed to solve very little (see Chapter 4). Without
the interest in organ procurement, *redefining* death appears gra-
tuitous.

Given Beecher's history of concern over ethical abuses in the
area of human experimentation, his lack of difficulty with
heart transplantation qua experimentation is noteworthy. In
"Scarce Resources and Medical Advancement," Beecher seemed
simply to bypass the discussion. Not human experimentation
and not established therapy, Beecher's unexamined characterization of heart transplantation registers his unquestioned enthusiasm for the procedure and his faith in its ultimate success:

"It is a therapeutic effort that will be widely practiced, once the rejection phenomenon is overcome." Acknowledging that "a considerable debate is at present under way concerning whether or not further heart transplants should be attempted until the rejection phenomenon in general is better understood and better controlled," Beecher offered no strong personal objection.[69] Feeling this way, he would not necessarily have seen the contradiction in his Harvard committee's having invoked a concern for relieving the suffering of the dying, while at the same time facilitating an effort to prolong it.

The minutes from the death and dying group are not extant in the Hastings Center archives. The correspondence between Callahan and Beecher, however, suggests that tensions between Beecher and members of the death and dying task force (specifically, Ramsey and Jonas and, to a lesser extent, William F. May) continued into the next year. The letters do not specifically mention the redefinition-of-death/organ transplantation controversy, but their timing strongly suggests that this was the main conflict in question.

"It seems to me that this whole controversy verges on the ridiculous," Beecher confided. "Some time I would like to go over with you the absurd misunderstanding that characterizes so much of what Ramsey has had to say. I throw up my hands when it comes to Jonas and to some extent in the case of May as well." Callahan attempted to appease Beecher. "You may attribute more strength and influence to Ramsey and Jonas than is the case," he suggested.

> My own sense of things is that there is certainly some argument and disagreement there, but that things are not nearly so polarized as the stories in the papers would make out. . . . In any case, it would make me very happy to see the death group get on to some other issues, and leave that controversy behind. I really don't like to see this kind of unpleasantness in any of our groups, since I think it sours the atmosphere in general.

Beecher replied that Callahan was "quite right"; there was tension in the group. "How could it be otherwise," he asked rhetorically, "when Ramsey makes statements in print"? He then made reference to Ramsey's criticisms in *The Patient as Person.*[70]

Then, in April 1971, Beecher offered to disassociate himself from the Hastings Center. "I have been taking an inventory of myself, vis-à-vis the group," he began. "I am struggling for honesty and clarity. My own contributions are so very different from everybody else's in the group, you and I must decide whether or not I am more of a liability to the group than an asset." Beecher then presented an inventory of his efforts regarding ethics and human experimentation, which he described as a summary "of my own situation as I see my very different response" (supposedly different from the death and dying group). He also indicated that "our group too much talks words when we should talk things." He concluded by allowing Callahan to decide whether Beecher should remain: "Well, I have tried to set down my strengths and weaknesses as I see them, and would like to abide by whatever you think is for the good of The Institute. I think you have done a remarkable job and I am proud to have been in on the beginnings of it. Whether I should continue with my 'off-beat' attitudes and activities I leave to you."[71]

Callahan defended the death and dying group's autonomy and integrity at the same time that he petitioned Beecher to remain. "My own policy toward the different groups," he revealed, "has been to steer clear of trying to tell them what they should be doing, other than, from time to time, reminding them all of the general Institute goals and rationale." He added that, "with the exception of yourself, I have had quite glowing reports of the discussions of the death group." Callahan's advice was for Beecher not to feel bound to the death group but to "take part only in those activities which interest you, skipping the rest." "I'd hate to see you drop out altogether," he affirmed. "I value your presence in the Institute, as one of the real and important pioneers in the field of medical ethics," Callahan continued. "We would hope to call upon you again, and perhaps more frequently," Callahan assured. He de-

clared his hope that Beecher "would make a distinction between the death group and the Institute as a whole, and not let
any misgivings about the former influence your feelings toward the latter." Beecher's response was a brief note thanking
Callahan for his "sensitive and thoughtful letter" and assuring
the institute's founder that he had "no wish to abandon the
Hastings Center. It was only that I felt guilty in not being able
to make any real contribution to the present task. I felt I was
keeping someone out who could have made a useful contribution."[72]

In July 1972 Beecher and other proponents of transplantation research quietly won an important cultural victory. Despite the early misgivings of some of its members, the Hastings Center group on death and dying gave its imprimatur both
to the Harvard criteria for irreversible coma and to the "official" story of the criteria's genesis, namely, that resuscitative
technology had prompted the contemporary need for reevalu-
ation.[73] The sanction came with the July 1972 publication in
JAMA of "Refinements in Criteria for the Determination of
Death: An Appraisal." The group statement mentions in pass-
ing that included in the category of those many uneasy people
were "a few members of this task force." There is, however, no
dissenting opinion offered. Hans Jonas's name is notably absent
from the list of those included on the Hastings Center death
and dying task force. "We can see no medical, logical or moral
objection," the report stated, "to the criteria as set forth in the
Harvard Committee Report." The task force drew this conclusion despite its disclosure that "frequent mention of organ
transplantation in connection with proposals offering new criteria of death has created an uneasiness on the part of many
people ... that the need for organs for transplant has influenced, or might sometime in the future influence, the criteria
and procedures actually proposed for determining that death
has occurred."[74]

The task force's opening statement strongly suggested that
advances in technology (as opposed to transplantation research
per se) had been the impetus for redefining death.

The growing powers of medicine to combat disease and to prolong life have brought longer, healthier lives to many....They
have also brought new and difficult problems ... An important
example is the problem of determining whether and when a person has died—a determination that is sometimes made difficult
as a direct result of new technological powers to sustain the signs
of life in the severely ill and injured.

   ... In a small but growing number of cases, technological intervention has rendered insufficient ... traditional signs as signs
of continuing life.

Still, the group professed that it could not deny the fact that
"the growth of the practice of transplantation with cadaveric
organs provided a powerful stimulus to reassess the criteria for
determining death." The "choice of the criteria for pronouncing a man dead ought to be completely independent of
whether or not he is a potential donor of organs." The appraisal
concluded, without explanation, that there were ample reasons, "independent of the needs of potential transplant recip-
ients ... for clarifying and refining the criteria and procedures
for pronouncing a man dead." This being the case, "the widespread adoption and use of the Harvard Committee's (or similar) clearly defined criteria will, in fact, allay public fears of
possible arbitrary or mischievous practices on the part of some
physicians."[75]

The Hastings Center approved of the Harvard criteria hoping, in part, to mollify public fears over harvesting organs from
brain-injured patients.[76] It did not, however, publicly address
whether criteria that had come into existence chiefly to foster
organ transplantation were adequate for determining when
and how to terminate treatment in order to end suffering. In
fact, the inadequacy of the Harvard criteria as a set of guidelines for terminating life-support measures in order not to prolong suffering makes the criteria understandable *only* as a
device promulgated to assist in the procural of organs for transplantation.[77]

The analysis here differs from that offered by historian
David Rothman in several ways. First, Rothman does not ex-

plicate the contradictory implications of the committee's ranked motivations for promulgating brain death criteria.[78]

Second, he takes seriously the notion that death was redefined, in part, to meet the problems created by artificial respiration—something that he considers a new technology. My analysis challenges both the idea that the technology was new and the notion that it necessitated a redefinition of death. The questions asked of extant data make a difference to understanding how bioethics took root and why it flourished. Rothman's book, being an account of a broad transformation in medical decision making, takes bioethics as a given and sees it as part of the 1960s attack on professional authority. In so doing, however, the ways in which the medical profession developed strategies to counter the broad attack on its authority—in part by harnessing the forces then coalescing to form bioethics—are overlooked. Focusing on the redefinition of death as part of the answer to how and why bioethics became institutionalized, the issue of whether problems in resuscitative measures were motivational in causing the Harvard committee to redefine death becomes crucial. The relevant question is not simply whether mechanical ventilation had ever caused physicians ethical unrest; it had. Instead, the relevant question is, What was the immediate cause motivating physicians to establish brain death criteria and to attempt to codify such criteria into becoming standard medical practice? Rothman makes clear that heart transplantation was the principal reason motivating the committee to redefine death. Left unexamined, however, is the question as to whether it was also true that ethical problems with mechanical respiration had been motivational as well.

It is clear that mechanical ventilation, along with many other devices and procedures, had given physicians ethical pause. We saw this most clearly in 1957 when anesthesiologists sought papal advice. Publicly codifying a new definition of death was not, however, a necessary outcome or the only solution to ethical anxiety over inappropriate use of the respirator. For example, training doctors to respect familial wishes and religious beliefs, and to familiarize themselves with the distinction between ordinary and "extraordinary" treatment and the principle of "double effect" were alternatives. In asking extant data to yield answers to the second question, however, we see that the immediate cause that motivated physicians to establish brain death criteria and to attempt to codify them into becoming medical custom was that such a move provided something that courts and legislatures could eventually validate as "legitimate" under the law, thereby relieving transplant surgeons from potential criminal liability. In this light, the committee's invocation of problems stemming from artificial respiration as motivational was misleading, and effective in the way it misled. This view is consistent with Rothman's data but is not buttressed by his narrative.

Third, Rothman believes that the Harvard brain death committee and its report represented "an important transition in the history of who shall rule medicine." The analysis offered here does not so much challenge this notion as question the extent to which it is true. Rothman correctly points out that the committee's "approach remained traditional in the sense that its members supported prerogatives for medicine that were in the process of being challenged, attempting, unsuccessfully, to make the definition of death a strictly medical concern." He also implies, however, that the move to include nonphysicians was an "unusual" step, one that represented a real shift in who ruled medicine. As we have seen, seeking counsel from nonmedical personnel was not a new idea: requesting advice from religious figures (and eventually from psychologists, sociologists, and philosophers) had been a motif since at least the fifties. Inviting a lawyer was the new feature. Along with other developments, this feature represented the real change, namely, the evolution of a strategy of submitting to limited legal and public oversight in an effort to preserve ultimate political authority.

The growth of bioethics does not represent a genuine shift in who rules medicine. It represents the profession's endeavor to limit potential threats to its ultimate control. It is a testimony to the success of this strategy, in this instance, that even in the midst of a raging controversy over human experimentation generally, the public never prohibited organ transplantation on this score; in fact, transplantation was facilitated by popular and legal support of new criteria for establishing when someone has died. The profession lost some prerogative in that it had to open its doors to public scrutiny on the issue of how death is determined. In doing so, however, it retained far greater discretion than it gave away. Calls for a moratorium had come from within the profession; heart transplantation was never barred by legal or popular sources. Heart transplantation research weathered its morally questionable experimental stage; the public allowed it to continue without turning its practitioners into "surgical criminals."

The Hastings Center approved a set of criteria that, ultimately, did little to clarify problems in diagnosing death and also left the way open for misapplication and tragedy. The brain death criteria, for example, wrought havoc for Karen Ann Quinlan and her family. Yet the notoriety the Quinlan case achieved would help to secure the rise of bioethics.

Chapter Four

"Sleeping Beauty"

KAREN ANN QUINLAN
AND THE RISE OF BIOETHICS
IN AMERICA

In the summer of 1974, while driving home from visiting with friends, Karen Quinlan and her boyfriend professed their love for each other.[1] The moment soon was marred, however, when Karen, given to making predictions, shared a premonition with her companion. The forecast was disturbing. "I'm going to die young," she said, "I'm going to go down in history." Karen already had offered this disquieting prophecy to two other friends the previous summer. On April 15, 1975, less than a year after Karen shared her thoughts with her boyfriend, she was admitted to New Jersey's Newton Memorial Hospital in a coma of unclear origin. On September 12, after twenty-one-year-old Karen's doctors refused her parents' request that their still comatose daughter be removed from the mechanical ventilator that assisted her breathing, Karen's parents began legal proceedings. Karen had since been transferred to St. Clare's Hospital, where her parents were told that because Karen was an adult, Mr. Quinlan would have to be appointed his daughter's legal guardian. On November 10, 1975, New Jersey Superior Court Judge Robert Muir Jr. denied the Quinlans' request to end the respiratory support and decided in favor of Karen's doctors: Karen would remain on the respirator. The Quinlans appealed to the state's highest court and on March 31, 1976, the New Jersey Supreme Court handed down *In Re Quinlan:* Karen's "right to privacy" now included her right to be removed from the machine. These events, and the public discourse they engendered, assured Karen a role in history; her prediction had come true.

The Quinlan tragedy played a significant role in the rise of bioethics.[2] A prevailing interpretation views the controversial case as spurring the "right-to-die" movement. The hospital's refusal to remove Karen's "vegetative" body from the respirator that breathed for her galvanized the public. Prior to Karen's coma there was only a low-grade awareness of a need to set limits in "saving" patients when prolonging dying was the only likely outcome. Among the more tangible immediate results of the post-Quinlan right-to-die movement were the "living wills" and redefinition of death statutes adopted by state legislatures across the country. In addition, ethics committees, today a feature of most hospitals, were first recommended by the New Jersey Supreme Court in *In Re Quinlan*.

Beyond this view of the Quinlan case, however, Karen's tragedy and its consequences can shed light on how common understanding is socially constructed. The nation responded passionately to Karen's misfortune. Her youth, gender, and comeliness, the mysterious origin of her unconsciousness, and the grotesque condition in which she was left gave Karen's tragedy a romantic, mythic quality. Karen's public existence, hovering somewhere between life and death, struck a cultural taproot of dread. Her enigmatic high school picture became a ubiquitous image—an icon for mistrust of professional authority, apprehension about technology, and fear of a "living death."[3] The discussion that accompanied her image reveals that public understanding, as well as misunderstanding, was full of doubt about authority and anxiety about technology.

Early news coverage of Karen's misfortune reported information that she was brain dead and was being kept "alive" artificially—a misconception that proved to be persistent. Although comatose, Karen was not "brain dead." She did not have a flat electroencephalogram, she responded to external stimuli, and she had sleep and wake cycles. Every doctor who examined Karen agreed that she was alive. But her condition was generally thought to be irreversible and "vegetative." None of the doctors who examined Karen believed that she would live long after being removed from the ventilator. In re-

fusing to turn off Karen's respirator, doctors were not attempting to keep a "brain dead" individual attached to a mechanical ventilator.

The Quinlan doctors' refusal to discontinue use of the respirator, and the litigation that ensued, created the impression
that such refusal was typical of medical professionals. Again,
this was not the case. Keeping "chronically vegetative" patients attached to ventilators was not a stated policy of organized medicine, nor was it customary medical practice. In fact,
physicians had frequently "pulled the plug" on patients like
Karen Ann years before 1975.

Early press coverage of Karen's tragedy and the ensuing litigation left the impression that the respirator and its overzealous use were novel causes of a new dilemma. Once again, this
impression was erroneous. Karen was less the victim of new
technology than of two new aspects of medical politics: the
1968 redefinition of death and the rising fears over the potential criminalization of medical practices. Public commentary
also revealed a readiness to believe that the New Jersey
Supreme Court had finally delivered Karen from the jaws of
technology. It was often assumed that the court had champi-
oned patients' rights against illegitimate medical authority,
that it had told the Quinlans that they could have their daughter removed from the respirator. In fact, the court permitted
the Quinlans to ask the doctors to remove her. The court's message to Karen's doctors was that they might remove her from
the device, but they were not required to do so. While the court
verbally affirmed Karen's right to privacy—a right it construed as sufficiently broad to include a right to be removed
from the respirator—the legal impact of the case actually was
to reinforce medical discretion on this score. Careful examination of the court's language reveals that what the law gave with
one hand in recognizing Karen's right to privacy, it took back
with the other hand in rendering the specifics of its declaratory relief. *In Re Quinlan*, often seen as the quintessential example of how external rules were brought to bear on the profession, was actually less a gift of freedom from technology to

patients like Karen Quinlan than it was a gift of freedom from liability to organized medicine.

The persistent nature of these misconceptions suggests that the alarm over a technology out of control and concern about a medical profession out of touch were not simply responses to the Quinlan affair; they had, in part, created it. A cultural predisposition of fear and mistrust was instrumental in giving birth to these misperceptions and further fueled the anxiety felt over the fate of Karen Quinlan and its social implications.

The Quinlan case also provides a window on the social ramifications of the politics of medical research. The anguish of the Quinlan family in the mid-1970s was the purchase price of organ transplantation experimentation in the late 1960s; research medicine's efforts to legitimize organ transplantation experimentation in the late sixties led to the establishment of the brain death criteria, a development that left the door open to the kind of tragedy visited upon Karen and her family. In 1968 proponents of organ transplantation research sought to legitimize a definition of death that would facilitate acquisition and preservation of human organs, namely, the "brain death" criteria (see Chapter 3). The traditional criteria for determining death, the cessation of respiration and heart function, effectively prohibited the "harvesting" of transplantable organs from a patient on a respirator whose heart and lungs were being artificially maintained through mechanical ventilation. The new criteria for establishing death, which focused on the cessation of brain activity, allowed transplantation surgeons to remove organs from such a patient even while that patient's lungs and heart were being kept operational through mechanical assistance.

Focusing on the needs of transplantation research, the authors of "A Definition of Irreversible Coma" (the brain death criteria) were not interested principally in creating standards for when and how to end human suffering. Moreover, little thought was given to the consequences of publicly seeking legitimization of such narrowly drawn criteria. The brain death criteria empowered doctors to turn off the respirators of potential organ donors with impunity; but they also contributed to the unease of physicians who would seek to discontinue the treatment of hopeless patients like Karen—patients who fell beyond the pale of the new definition because they still possessed some brain activity. *In Re Quinlan* made it clear to doctors that they would not be prosecuted for "pulling the plug" on non-brain-dead, chronically vegetative patients. In this sense, the Quinlan case represents the success of medical research's controversial effort, begun in 1968, to win acceptance of the criteria for irreversible coma or "brain death."

The court abetted the mistaken impression that refusing to disconnect respirators was standard medical practice. Although some physicians were candid enough to admit to journalists or to publish in the more obscure professional journals that "pulling the plug" was customary, it was quite another matter to admit as much in a court of law. Under the "white light of litigation," doctors would not acknowledge that turning off life-support systems in hopeless cases was standard medical behavior.[4] The legal process concealed the nuanced reality of medical practice. As a result, the court had little choice but tacitly to confirm the notion that physicians routinely submitted patients in vegetative states to slow, horrendous decay.

Finally, the Quinlan case provides a window on the establishment and growth of bioethics as a field of "expertise." When the criteria for brain death were promulgated in 1968, demand for bioethical services came most urgently from backers of transplantation research. Supporters of this research wanted an "objective" source to influence popular opinion toward accepting a definition of death that would facilitate transplantation experimentation (see Chapter 3). In 1975, when the case of Karen Quinlan gripped the American public, widespread fear fueled the demand for bioethics. This fear was expressed through the press, the courts, legislatures, and bioethical analysis, as Americans searched for ways to understand and control what they perceived as the threat of technology to the dignity of human life and death. The kinds of answers bioethicists offered insured their own continuing viability—

by confirming the threat of technology and the virtue of remaining vigilant.

Despite the intense focus on technology, Karen's fate and the
dilemma it brought to light were not the results of unique circumstances accompanying the use of mechanical ventilation.
What happened to Karen after she fell into unconsciousness
was a historic moment of public crisis, a creation of the social
and political milieu of the 1970s. The elements of this milieu
included medical professional anxieties over potential criminal
liability, the tense atmosphere engendered by the medical malpractice "crisis," the increasing scrutiny of medical research
and practice by the press, a strong cultural animus of fear and
mistrust of technology and professions, and the ethical par-
lance developed by bioethicists.

The Quinlan controversy offers an example of how, during
the seventies, the emblematic radical political critique of the
sixties was transformed. Throughout the 1960s, challenges to
technology and authority were explicitly political and threatened the legitimacy of the infrastructure of medicine, science,
and law (see Chapter 1). By the mid-1970s the radical nature of
these challenges was ebbing. The seventies saw the channel-
ing of such confrontations through the individualizing legal
process and into discussions of the moral nature of social issues.
The resulting solutions to the perceived social threat posed
by Karen's situation—legal dispositions and ethical delibera-
tions—suggest that radical critique was being absorbed into
existing institutions. Discourse was in the process of sifting
explicit considerations of power into discussions on law and
ethics. Turning now to the national reaction to Karen's predicament, this chapter brings these developments into focus.

News coverage began soon after Joseph Quinlan petitioned
for guardianship of his adult daughter. Three days after the
Quinlans' lawyer filed the complaint at the Morris County
Courthouse on September 12, 1975, the Quinlan family's front
lawn was covered with reporters, a condition that did not abate
until resolution of the controversy the next year. Later that

first week, correspondents from NBC and CBS arrived at the Quinlan home ready with camera crews. Before the end of 1976, when the New Jersey Supreme Court rendered its decision, the story of the "girl in coma" led to countless newspaper articles in the United States and around the world, magazine exposés, news conferences, two full-length books, a television documentary, and a made-for-television movie. The Quinlans were receiving hundreds of phone calls on a daily basis. One reporter offered the family $10,000 for a photograph of Karen in her bed. Another caller, claiming to represent a worldwide news-photo agency, offered $100,000 for a photograph.[5] Lawyers on both sides of the case, the judge, and the hospital were inundated with phone calls from the news-hungry public. Denville, New Jersey's St. Clare's Hospital, where Karen had been transferred, hired a public relations expert to deal with the situation. The hospital also posted round the clock guards outside the intensive care unit where Karen stayed to prevent faith healers and the curious from entering.

The impassioned national response to Karen's misfortune was due, in part, to repeated press characterizations of her coma as "mysterious." At a party the night of April 14, Karen, ill and fainting, was taken home early by friends. Later, when a housemate checked on her, he found that she had stopped breathing. Resuscitative efforts were of no avail; her brain had been deprived of oxygen for too long. Tests found prescription levels of valium in her blood along with traces of quinine consistent with having ingested gin and tonic. The amounts were not sufficient to cause serious brain damage, however, and the combination of drugs and alcohol was never confirmed as the source of the problem. Thomas French, who knew Karen and was with her during the last days of her life, described her as "self-destructive, popping whatever pills she could get her hands on and often drinking to excess."[6] A roommate of Karen's, Robin Croft, acceded that occasionally they "tied one on" and that Karen "might have taken a few pills for a high," but she insisted that Karen "wasn't into drugs," and denied that they had been living a life of drugs and drink.[7] Newspapers scarcely reported these renditions, and accounts that associated Karen with an unwholesome life-style faded quickly. The Quinlans' lawyer denied that alcohol and drugs had caused Karen's "illness" and maintained that any such combination had been dismissed as causes of the coma. After an investigation, the Sussex County prosecutor concluded that the twenty-one-year-old woman was an "innocent" victim.[8]

Doctors also discovered bruises on Karen's body and a bump on her head.[9] Despite Mrs. Quinlan's explanation that her daughter had fallen down a flight of stairs days before attending the party, Karen's injuries were considered "unexplained." After a grand jury probe, State Attorney General Hyland announced that the investigation had "failed to establish that foul play was the cause of the coma."[10] But what was the cause? Karen remained the perfect victim for this unfolding passion play—innocent and indecipherable.

Efforts to discern who Karen Quinlan had been produced an incomplete portrait of an appealing young woman. News coverage, commentary, and quickly emerging biographies emphasized the youth, wholesomeness, and former vitality of the nation's "sleeping beauty."[11] Karen was the "21-year-old ... who loved skiing and snowmobiling." "Karen had such a love of life that she would not want to go on living [attached to a machine]," according to her mother.[12] One essayist, concerned about the beguiling effect of such characterizations, cautioned that Karen's youth and beauty should not distract American society from the fact that "in our hospitals there are thousands [like her]."[13] Ironically, despite the exotic nature of Karen's circumstances, a disquieting sensibility persisted: what had happened to her could happen to anyone. Seen in this light, Karen was "a typical girl-next-door, neither beauty queen nor plain, neither 'intellectual' nor 'dumb,' neither wild nor conservative."[14] Karen was an Everywoman demonstrating to the world what it meant to be "dying in the age of eternal life."[15]

Benign images of Karen before her coma contrasted starkly with the lurid imagery of her grotesque condition attached to the machine. Courtroom testimony of expert witnesses pro-

vided the earliest reports of her physical state. Eventually, journalist B. D. Colen became one of very few nonmedical personnel allowed to see Karen. His account of her deterioration
combined the purely descriptive with dramatic analogies suggesting the idea of science torturing nature:

> " ... her strong athletic limbs wither[ed] to skin-covered sticks
> that were drawn against her body and frozen by calcium deposits
> so they would never move again. She looked like a bizarre pray-
> ing mantis, with a respirator tube protruding from her throat as
> a pin would hold an insect specimen in a collection box. . . . She
> moved, frequently, but only in the mindless way in which Priest-
> ly's frogs moved. She moaned, but only because the involuntary
> movements of her vocal cords played tricks with the air her res-
> pirator and lungs had forced through her throat. Her eyelids
> would flutter and open, and her eyes roll from side to side, but one
> would roll left as the other rolled right.[16]

From sleeping beauty to the girl-next-door to the grotesque aspect of an insect tortured by technology, competing images of
Karen's tragedy shaped public awareness into shifting realities
and durable myths.

Support for the Quinlan family's decision to remove Karen
from the ventilator eventually dominated public discussion.
Some commentators were, however, disturbed by the unseemly
request to have a child's life deliberately ended. "These people
are well meaning," Karen's father responded, "but they don't
know the case—how hopeless it is."[17] Thus, Charles McCabe,
columnist for the *San Francisco Chronicle,* characterized the
dilemma as a matter of two issues, euthanasia and free will,
Karen's will. "There is no really solid evidence," he admonished, "that the patient wills her own death."[18]

News reports summarized Karen's court-appointed attorney's assertion that "her medical condition is not hopeless and
the petition by Joseph Quinlan (to have his daughter's respirator removed) should be denied until all chances of a cure have
been exhausted."[19] Headlines betrayed biases: "Parents Ask
Court to End Woman's Life," and "Parents Seek Death of Girl

in Coma," were succinct spins implying harsh criticisms of the Quinlans and their request.[20]

Millions of Americans read that lawyers representing Karen's imputed interests (separate from her father's) construed removal of the respirator as an act of euthanasia. Attorneys hoped to evoke a shock that may be difficult to gauge given the current impetus to legitimate physician-assisted suicide and euthanasia under the "right-to-die" banner. But in 1975 the concept of euthanasia could provoke references to "Nazi atrocities" as easily as "death with dignity." "I've heard 'death with dignity,' 'self-determination,' 'religious freedom,'" one lawyer began, "and I consider that to be a complete shell game that's being played here. This is euthanasia." "You can't just extinguish somebody because she's an eyesore," he proclaimed.[21] The doctors' attorney exhorted the court that the Quinlans' suit was analogous to the Nazi atrocities and gas chambers of World War II. Where could one draw the line, he asked, between a person whose life is worth saving and one whose existence is so futile that life should be ended.[22] He warned dramatically that, should the Quinlans' petition be granted, the court would "open the gates to the death of thousands of people in the United States who may have a low quality of life." Invoking religious authority, he queried, "Dare we defy the divine commandment, 'Thou shalt not kill'?"[23] Hoping to chasten the court, he placed the litigation in chilling legal perspective: "It is the first time in the long history of Anglo Saxon jurisprudence that a human being, universally recognized as being alive legally and medically, may be put to death as the result of a civil suit."[24]

The judge—and the public—were told, "You are being asked to place your stamp on an act of euthanasia."[25] One *Chicago Tribune* reader felt that, in keeping Karen on the respirator, society was championing the weakest members of society:

> One of the great philosophical errors of the day is the seemingly enlightened view . . . that the worth and humanness of a man lies

exclusively in his social utility.... It is pressing the medical profession into what it euphemistically regards as a service to man—
the murderous eradication of useless people.... Be thankful for
the good physicians who are speaking out in favor of life and run
from anyone who encourages death with dignity.[26]

Time magazine argued that should the court approve removal
of the respirator the effects might be far-reaching: if carried
to its logical conclusion, such a decision might free thousands
of people to seek legally to end the lives of suffering wards.
Such a decision could

> be applied in state hospitals, institutions for the mentally retarded
> and for the elderly. Such places currently house thousands of peo-
> ple who have neither hope nor prospects of a life that even ap-
> proaches normality. A decision to remove Karen's life-support sys-
> tem could prompt new suits by parents seeking to end the agony
> of incurably afflicted children, or by children seeking to shorten
> the suffering of aged and terminally ill parents.[27]

Some feared the Quinlans' request was, indeed, a step on the
slippery slope to state-approved euthanasia. Others objected to
the petition because their personal involvement with recovered
comatose victims led them to believe that Karen, too, might one
day be restored. A Salem, West Virginia, woman's experience,
for example, led her to disagree with Joseph and Julia Quinlan:
"Though I have great sympathy for the Quinlan family, especially Karen Ann, I do not agree with them. I'm for holding on
to every shred of life. My husband was in a coma in 1958....
He's fine now—and I wish as much for Karen."[28] The opposition of still others, influenced by religious beliefs, was a matter
of faith. "God has given me faith to believe that your daughter
Karen can be healed," one Cedar Rapids, Iowa woman wrote the
Quinlans, "in spite of what the doctors say."[29]

Where early opinion may have been split evenly between
those who supported and those who opposed her continuation
on the respirator, eventually the majority of those expressing
opinions backed the Quinlans' plea to remove the respirator. In
October 1975, before the trial, St. Clare's Hospital reported that

the two hundred letters it had received were all opposed to shutting down the machine. The Quinlans, however, reported that although the mail had been split about fifty-fifty up to the time of the trial, since the trial's beginning, the mail was running two-to-one in favor of letting Karen "die with dignity." Mr. Quinlan speculated that the shift in public opinion was due, most likely, to descriptions of his daughter's degenerating condition.[30]

"Sleeping Beauty is in a coma," began *New York* magazine's account in early October. "Beautiful, long-haired, 21-year-old Karen Ann Quinlan has been in a coma for five months."[31] But later, as the trial unfolded, news reports combined romanticized references to Karen's "sleep" with graphic descriptions of her condition. *Newsweek*'s lengthy November exposé featured an artist's depiction of "the girl in a coma." It featured a slim-faced, peacefully reclining young woman with long dark hair; surely this figure could one day be roused from her serene slumber. The same article, however, also quoted Dr. Julius Korein who, when asked in court to characterize Karen's mental age, responded that Karen's situation made her equivalent to "the anencephalic monster. . . . If you put a flashlight in back of the head the light comes out the pupils. They have no brain."[32] Ethereal images of the world's most famous patient could not long be sustained in the light of competing references to her deformed "fetal" posture and her "irreversible," "chronic vegetative" state. Reports suggest that the public came to feel, with the editor of *America*, that, "at some point medical technology ceases to serve human ends. Every description of Karen Ann Quinlan's condition argues that she has reached that juncture."[33]

Early news coverage jostled between chilly references to Karen's "adoptive parents" and more sympathetic portrayals of Joseph and Julia Quinlan as "her anguished" or "loving parents."[34] Published excerpts of trial testimony allowed the public to hear the Quinlans speak in their own voices. The abstract "father seeking guardianship of his adult comatose daughter," yielded to descriptions of a long-suffering parent sincerely seeking to help his child. "This poor man," suggested one op-ed writer, "with his obvious sincerity and basic decency may well go down as one of our modern folk heroes."[35] Mrs. Quinlan was simply "a mother trying with love and concern to have hospital machines turned off so that her hopelessly ill daughter can die in peace and dignity."[36] The "unsophisticated nobility of her parents" was persuasive.[37] "After hearing Mr. Quinlan's articulate testimony," wrote one woman, "I was so moved by his sincerity in 'turning our Karen's life to the loving Lord,' I don't see how anyone can deny you this favor."[38]

Some members of the public, unconvinced early on of the virtue of the Quinlans' motivations in seeking to have their daughter's life support terminated, changed their views after reading what are generally termed human interest accounts of the tragedy. The Quinlans' lawyer noted how the news-hungry swarm of reporters always immediately quieted down when Mr. and Mrs. Quinlan approached to speak. He interpreted the crowd's transformation as a sign of its reverence: "As always happened, a normally unwieldy professional throng became respectful and even reverent when Joe and Julie arrived. Their collective sympathy was always apparent."[39] Actress Shirley MacLaine quickly learned just how deeply the public had come to identify with Karen and her parents when, on her opening night at the Palace Theatre in New York, she called New York City "the Karen Quinlan of American cities" to indicate New York's deteriorated condition. The audience met her comment with boos and catcalls. MacLaine felt compelled to call Joseph and Julia Quinlan to apologize for her insensitivity.[40]

The shift in public discourse from some disapproval to support for the Quinlans' request was not unanimous. On November 10, 1975, New Jersey Superior Court Judge Robert Muir decided that, if doctors felt that Karen should remain on the ventilator, it was not up to the courts to contravene their judgment. A minority public view supported this decision. Similarly, when the New Jersey Supreme Court reversed Judge Muir's decision and ruled that Karen could be removed from the respirator, those in disagreement with the judgment made their displeasure known. "If awareness is to be the criterion for legal survival in this country," wrote one Wisconsin resident, "then we all are at some time or other in our lives going to be in grave danger."[41] Later, an organization called the Human Life Amendment Group opposed the New Jersey Supreme Court's ruling and attempted, unsuccessfully, to get the United States Supreme Court to review the case.[42]

Columnist George Will agreed that Judge Muir's decision was the correct one. "Society's consolation in this sad case," Will argued, "is that the law's protection of life has been affirmed."[43] An editorial written for the *Los Angeles Times* by a physician reported that her black colleague had breathed a sigh of relief when the lower court ruling was handed down. Why?

> All of the powerless in American society can take heart from the ruling. . . . In showing respect for Miss Quinlan's life, he [Judge Muir] affirmed everyone's right to life—what ever the circumstances of a particular individual. . . . Whenever the value of life is diminished, those least able to defend themselves—the very young, the very old, the poor, the sick, the mentally disadvantaged and racial or religious minorities suffer the most.[44]

Ultimately, those who favored keeping Karen on the machine were but dissenters against a wave of opinion in favor of releasing Karen from the machine. When the lower court declined to overrule doctors' orders and shut down the ventilator, the *New York Times* was one voice among many rebuking the court or expressing regret over the decision.[45] It chided Judge Muir's "excessive conservatism" in refusing "the opportunity to help bring common law into step with the amazing new capabilities of medical science." The editor's scolding tone, in line with public opinion polls, registered disapproval at a decision that could only "prolong the agony of her devoted family," who must now "suffer the indefinitely continued anguish of knowing that this once vital young woman is merely a mass of metabolizing protoplasm kept 'alive' artificially."

Opinion polls confirmed public discussion favoring "pulling the plug." In March 1976, the month of the New Jersey Supreme Court's decision, the *Ladies Home Journal* reported the results of a sixteen-city survey of women on the topic of, "'Pulling the Plug' . . . Is It Murder or Mercy?" The majority of women polled "overwhelmingly" agreed with the Quinlans' decision to have their daughter removed from the respirator: 60 percent of the respondents believed that the Quinlans were right in fighting to have the respirator removed; 69 percent answered that they would want the respirators turned off if their loved ones were terminally ill. When asked what they would want done if they themselves were the terminally ill patient in question, 81 percent answered that they would have the machine discontinued. Furthermore, while many qualified their answers, 77 percent of those who expressed an opinion stated that someone who "pulled the plug" on a terminally ill patient should not be prosecuted.[46] A poll conducted by Lou Harris in 1977 (and again in 1981) offered clear evidence that public opinion dramatically favored families' prerogatives to have life support removed.[47]

Occasionally, letters and commentary would cite the financial cost to society of treating Karen, and people like her, as a reason for shutting down the respirator.[48] Karen's medical expenses were paid by state hospitalization insurance and there were those who questioned the wisdom and fairness of spending precious treasure on hopeless patients.[49] Utilitarian cost-benefit thinking, however, was muted beneath the more anxious concerns of mistrust of authority and fear of technology.

Mistrust of authority extended to lawyers and the legal process as well as to doctors and medical practice. "The thought that attorneys for a hospital and a deputy attorney general," began Dick Harbottle of California, "can insensitively supersede my wishes, or those of my loved ones, appalls me."[50] Lynn Tressler wrote the *Washington Post* that "the family must be the ultimate authority. I do not mean to suggest the legal process or medical professions should not play any role, but . . .

they should interfere as little as possible, and not impose their will or morality."[51] This general antiauthority attitude persisted throughout the long months of the Quinlan litigation and sometimes took on a sarcastic, hard-edged tone. After the lower court handed down its ruling, Donald Kirk of the *Chicago Tribune* wrote that Karen Quinlan now "had the right to keep on living in a coma as long as she can breathe with the aid of a respirator."[52]

Even before the municipal court found in favor of Karen's doctors, animosity toward the medical profession was evident.[53] The lower court's ruling favoring Karen's physicians only fanned the flames of contention. The press quoted Judge Muir's opinion favoring medical discretion at length: "the nature, extent and duration of care is the responsibility of the physician. What justification is there to remove it from the control of the medical profession and place it in the hands of the courts? . . . There is a duty to continue the life-assisting apparatus if within the treating physician's opinion it should be done."[54]

Those who either favored preserving the integrity of medical discretion or who believed that shutting off the machine was abandoning a defenseless patient could read such reasoning with equanimity.[55] But for those waiting for Karen to be released from a living hell, Judge Muir's opinion was a line drawn in the sand. "If Judge Muir is right," wrote Fordham Law School's Charles Whelan, "[patients in irreversible coma] become much more than their doctor's patients: they become their doctor's prisoners. . . . [Judge Muir] has misinterpreted the extent of American society's commitment to the doctors." A *Los Angeles Times* reader sarcastically prophesied that, "There will come a time when almost any 'fatal' anomaly of human physiology can be circumvented by connecting the body to various combinations of machinery. Then all of us— including Judge Muir—can look forward to the possibility of eventually being sustained as Karen for the rest of our 'natural lives.'"[56]

The lower court ruling intensified the hostility felt toward doctors. For one resident of Long Beach, California, doctors had no business making moral judgments over patients' lives: "It is extremely important that we perceive the danger here. Doctors are qualified and very knowledgeable in describing a patient's condition. . . . But to suggest that they are the ones who should determine if continued treatment is worthwhile is not correct. . . . A doctor's job is not to make moral judgments."[57] The *Chicago Tribune*'s stark logic emphasized the recalcitrance of Karen's physicians: "Although they agree there is no hope of . . . recovery," it related, "doctors have refused to discontinue the emergency measures."[58] Even the New Jersey Supreme Court's reversal of the lower court's decision did not quell public antagonism toward doctors. Columnist Ellen Goodman expressed her dismay with professional medicine, asking rhetorically, "Have we given up birth and death to medicine? What does it do to us when we treat death as a technical error?"[59] When B. D. Colen penned his account of the Quinlan case, his hostility toward the medical profession was evident: Karen's doctors "would not acknowledge their human shortcomings and insisted on playing God."[60] After the lower court rendered its opinion, instead of neutral headlines such as "Court Rules in Quinlan Case," major papers led with "Judge Rules Karen Must Be Kept Alive by Respirator" and "Quinlan Denied Control of Daughter's Respirator." Accentuating the court's coercion and refusal, these headlines thinly disguised support for the Quinlans in their struggle against medical control.[61]

The respirator and the specter of Karen Quinlan hovering between life and death haunted the American public. Public discussion revealed the widespread, uncontroverted perception that medical technology was threatening the dignity of human life and death as never before. The Quinlan family's young, idealistic lawyer helped set the tone for this discourse, believing the case to be one of humankind's battles against technology. "These poor people really need help," he told a colleague. "The whole world needs help. It's a man against technology."[62] Later, he asked the court rhetorically, "Can anything be more degrading than the concept that death can be cheated if we can only find the right combination of wires, tubes and transistors?" "We're coming to the court and saying, 'Help us resolve this problem where medical technology has outstripped the law.'"[63]

News reports, letters, and commentary responded in kind. Journalist B. D. Colen speculated that, should the court decide against the family, "our hospitals and nursing homes will soon become filled with machine sustained organisms devoid of all human qualities."[64] Physicians, typically, were more sanguine in their reaction to the case. President of the American Public Health Association, Dr. C. Arden Miller, however, waxed dramatic when he warned that if society always lets the doctor decide in these matters, "then we are surrendering humankind to the dominance of rampant technology."[65]

When the New Jersey Supreme Court reversed the lower court decision and ruled that Karen could be removed from the machine, those distressed by the earlier decision saw it as a deliverance from technology run amok.[66] Even after the New Jersey Supreme Court cleared the way to free Karen from her "prison" and coverage receded from the front pages and editorial sections of newspapers, the need to rein in a technology out of control continued framing discourse about the social and philosophical ramifications of the case. The editors of the *National Review*, for example, foresaw only increasing difficulties with medical technological "improvements": "No doubt, with the continued improvement of medical technology, we will face such issues with growing frequency. . . . Suppose the technology becomes available to keep a person in Karen's condition 'alive' forever? It is conceivable that with the development of advanced transplant techniques the very idea of individual identity will begin to dissolve. The possibilities, certainly, are mindboggling."[67]

Ruminating on the case, attorneys associated with the litigation characterized the social implications of the case luridly. In a coauthored reflection on the case, New Jersey's attorney general offered eerie images of hospital life and "phantom beings"—a new technological American gothic:

Today, in every hospital in the country, a shadowy community exists comprised of phantom beings tenaciously clinging to worldly existence. Sharing the attributes of both life and death, they form a new generic class of existence. The startling progress that has taken place in medical science and technology has obliterated the distinction between life and death and raised serious questions regarding the care and treatment of these unfortunate people.[68]

The year following the trial, the Morris County prosecutor who tried the case offered grim forecasts, similarly distressed by the possibility of hospitals that are "populated by beings without consciousness."

That medical technology has out-distanced law and traditional ethics is to state the obvious. The same techniques that have made possible the almost commonplace restarting and transplanting of hearts have also created the possibility—and all too often the fact—of a community of intensive care wards of modern hospitals populated by beings without consciousness, whose very "humanness" is subject to fervent debate.[69]

In his popular book on the Quinlan tragedy B. D. Colen wrote that Karen, in her comatose state attached to the respirator, was a twentieth-century Frankenstein monster, one of many across America:

Victor Frankenstein . . . is alive and well. . . . He is still bringing the dead back to life, and he continues to take organs and parts from one body to implant in, and graft onto, another. He uses jolts of electricity to restart stilled hearts. . . . His laboratory and operating theater are still filled with humming, buzzing pieces of exotic equipment and jars and vials of blood and other fluid. . . . there are thousands of Victor Frankensteins. . . . They are physicians to whom we give the battered corpses scraped off our highways and the bloodied victims of urban carnage—physicians who usually give us in return not monsters but our friends and relatives made whole once again. . . .

But . . . they sometimes create monsters. . . . They [the monsters] are the Karen Ann Quinlans who lie in intensive care units and nursing homes all across America.[70]

But if Karen was a monster, she was also a symbol. Karen and her parents had become "symbols of a deeply moving struggle against technology and the beneficence of medical science run amok."[71]

The most trenchant response to the case was the dread that what happened to Karen could happen to anyone.[72] One Ohio lawyer, contemplating Karen's plight, wrote to the Quinlans of her apprehension: "I have had brain surgery, my medical prognosis is one more year to live. I have informed my neurosurgeon, my family physicians, friends, family, priest and a Protestant minister that I do not wish to be kept 'alive' under the same circumstances."[73] Detroit resident M. Hathaway read *Newsweek*'s story on Karen "with sheer terror." "A perpetuated state of dying," she wrote, "is an inhuman and unearthly thing. It is outrageous that any human would sentence another to that type of hell."[74] Mistrustful of medical authority and haunted by misuse of technology, many sought concrete ways to help ensure that they would die as they saw fit.

A dramatic indication of public dread is the resurgence of interest in right-to-die legislation. Concern about the role of medical science in prolonging dying had been a low-grade social issue for the public and the medical profession for at least a decade (see Chapter 3). The Quinlan case now became the catalyst for a grass-roots effort to address the matter head-on. Healthy individuals began signing "living wills," specifying that, in the event that the signatory became terminally ill, no extraordinary means would be used to prolong dying. William F. Buckley Jr. signed his name to a version of this document, "A Directive to My Physician," and published it in his syndicated column. He counseled the public that this was an effective way of avoiding the "Quinlan Syndrome."[75] After Abigail Van Buren, the "Dear Abby" columnist, explained the living will in her column, the Euthanasia Education Council was swamped by sixty thousand inquiries about the will and requests for the document.[76] Between 1969 and 1975, the council distributed 750,000 copies of its rendition of the living will.

In the year and a half after the Quinlan story hit the presses, the council received 1.25 million requests for the text.[77]

Before 1975, the legislatures of five states had introduced right-to-die legislation. After New Jersey's Superior Court denied Karen's parents' request to overrule doctors' orders, seventeen bills were initiated; all were eventually defeated. By April 1977, however, thirty-eight legislatures had submitted fifty such bills, with California and Idaho turning them into laws.[78] Many citizens apparently agreed with Mrs. B. R. Wheeler of North Carolina who wrote that as a young woman influenced by "Karen's dilemma" she had taken "a long hard look at the 'right to die' issue" and suggested obtaining a "living will." "Personally," she said, "I sleep better at night knowing my beliefs and wishes are stated in black and white, should circumstances render me unable to speak for myself."[79] The Quinlan case's jolt to the right-to-die movement was enduring. By 1987 thirty-eight states and the District of Columbia had enacted living will statutes.[80] Today, the case is an important legal antecedent for those seeking legislative approval of physician-assisted suicide (see the Epilogue). When the public sought living wills as the antidote to the technological nightmare induced by Karen's tragedy, however, it was reacting to misconceptions about the nature of the Quinlan litigation.

The ready focus on the drama of perilous technology that prompted this unprecedented legislative activity obscured the more mundane sources of the Quinlan litigation and fostered continuing misconceptions regarding the nature of Karen's situation. The doctors' refusal to terminate treatment resulted in litigation less because of problematic technology than because of professional fears over criminal liability. The question of liability had been an issue from the beginning of litigation. It was addressed by the attorneys in pretrial conferences and, ultimately, by both New Jersey Superior and New Jersey Supreme Courts.[81]

Quinlan lawyer Paul Armstrong argued before the New Jersey Supreme Court in January 1976. The court's dialogue with

Armstrong clearly implicates fear of liability as motivational, as it cuts through a discussion of doctors' beliefs to their underlying concern about liability. The justices asked Armstrong if he wasn't "asking the Court to order the doctors to terminate the apparatus?" "Absolutely not," he replied. "Aren't you really asking us to overrule the doctors' decision?" the Justices persisted. No, responded Armstrong. If the doctors in question weren't comfortable turning off the respirator, the Quinlans would simply find other physicians who would do it. When the court pressed Armstrong as to why, then, he hadn't just already found other physicians who would do what the Quinlans desired, Armstrong explained that he hoped the court would provide "guidelines." Physicians wanted those guidelines, he believed, to help them make decisions to terminate life-supporting equipment. "Are you not also asking that if the physicians accede to the parents' request, it will be without any adverse consequences to those physicians?" the court probed. "Yes, to that," agreed Armstrong.[82]

The New Jersey Supreme Court ultimately professed its belief in the doctors' statement that fears over liability had not motivated them to keep Karen on the respirator. The court went on to affirm the importance of this concern, however, and characterized the issue of possible liability as a "brooding presence":

> The modern proliferation of substantial litigation and the less frequent but even more unnerving possibility of criminal sanctions would seem, for it is beyond human nature to suppose otherwise, to have bearing on the practice and standards as they exist. The brooding presence of such possible liability, it was testified here, had no part in the decision of the treating physicians. As did Judge Muir, we afford this testimony full credence. But we cannot believe that the stated factor has not had a strong influence on the standards.

After expressing hope that its opinion "might be serviceable to some degree in ameliorating the professional problems under discussion" (namely, liability), the New Jersey Supreme Court discussed the usefulness of ethics committees; and then it offered the medical profession the reassurance it needed: physicians did not have to fear criminal liability.

> The County Prosecutor and the Attorney General maintain that there would be criminal liability. . . . We conclude that there would be no criminal homicide in the circumstances of this case. . . .
> . . . The termination of treatment pursuant to the right of privacy is, within the limitations of this case, ipso facto lawful. . . .
> . . . the exercise of a constitutional right such as we have here found is protected from criminal prosecution.[83]

In Re Quinlan made it clear that doctors who removed chronically vegetative patients from ventilators would not be found guilty of murder.

The refusal to turn off the respirator was a decision made in the liability-vexed atmosphere of the beginning of the medical malpractice crisis. Organized medicine had declared a "crisis" because the increasing number of lawsuits brought against doctors by patients had driven up insurance costs.[84] In this tense climate, potential liability was certainly an issue for the medical community. The Quinlan incident, however, was particularly acute because it held out the possibility of criminal, not merely civil charges against physicians—a legal difficulty of a much greater order of magnitude. Two developments at the intersection of law and medicine made medical professional worries over criminal liability a live issue in the mid-1970s: the Harvard ad hoc committee's 1968 redefinition of death and the Edelin case, which occurred in February 1975.

The 1968 criteria for brain death were a narrowly construed set of guidelines. The chief concern at the time was to increase the supply of transplantable organs. If irreversibly comatose individuals were to be targeted as a source of harvestable organs, however, it was imperative to render guidelines that left as little room as possible for the removal of organs from people who might have even a modest chance of recovery.[85] (During the Quinlan trial none of the doctors who testified would say with absolute certainty that there was "no hope" for Karen.)[86]

Thus, before the machine could be turned off, requirements ensuring that the individual would never revive had to be met: a flat electroencephalogram over a twenty-four-hour period, no response to external stimulation, no spontaneous breathing, and pupils fixed and dilated. Karen met none of these requirements. If the chief concern in 1968 had been the establishment of criteria for the principled removal of patients from mechanical ventilation to prevent prolongation of suffering, the guidelines would have looked quite different—a broader set of rules to include cases of "chronically vegetative" patients. Indeed, it would not have been necessary to define irreversibly comatose patients as dead at all. The guidelines simply could have specified that should a patient meet the criteria, the respirator may be turned off—without a declaration of death. It became necessary to declare irreversibly comatose patients dead by these new criteria to avoid lawsuits from the relatives of patients whose organs had been removed while their bodies were still breathing (owing to mechanical ventilation).

In a 1969 article for the *Journal of the American Medical Association*, the chief author of the Harvard brain death criteria, Henry Beecher, appeared to be trying to rectify the damage done by having promulgated such narrowly drawn criteria for removal of life-support measures.

> It is to be emphasized that types of patients other than those exhibiting the brain death syndrome *may* also have supporting measures discontinued as in the following examples (this should be done with the concurrence of the family): (1) situations in which the physician in charge, supported by a neurologist or neurosurgeon, believes, on the basis of the evidence, that the brain is irrevocably destroyed, even though all criteria of the brain death syndrome are not present. . . .[87]

As it was, however, the guidelines themselves had been narrowly cast, excluding Karen and patients like her. As the 1968 "redefinition" of death became increasingly public and a matter of legislative validation, "pulling the plug" on potential organ donors became less problematic; doing the same for people like Karen, however, became a more risky proposition for physicians, not a less risky one. What had once been a matter of customary medical practice, the removal of chronically vegetative patients from respirators, could now be seen as unlawful killing under the law.

Harvard's 1968 codification of brain death was causing unanticipated hazards for physicians practicing in the litigious social environment of the 1970s. "Pulling the plug" on chronically vegetative but not brain-dead patients could now more easily be interpreted as a criminal act. In 1974 Dr. William Gill was clinical director of the Shock Trauma Unit for the Maryland Institute of Emergency Medicine. Dr. Gill's uncommon candor suggests the chilling effect this situation had on medical practice. "Any day of the week," he explained, "we will have two such patients . . . [who are being allowed to die]. . . . The majority we are still very cautious with and really wait longer than we should—basically wait until the patient is brain dead. If you wait long enough with a hopeless patient on a respirator most of them will become brain dead, and then you have a definition of death."[88] Thus, when the Quinlan dilemma was taken beyond the hospital corridors, the new arena was neither a church pulpit nor the halls of academe; only a court of law could render ultimate resolution.

The possibility of criminalizing medical practice was driven home to the medical community by the fate of Boston obstetrician Dr. Kenneth Edelin. When news of Karen's tragedy and the ensuing dilemma hit the front pages of newspapers across the country, the medical profession was still reeling from the legal implications of the Edelin case. Dr. Edelin, coordinator of Ambulatory Services at Boston City Hospital, was charged with allowing an infant to die after performing a late-pregnancy abortion. The trial produced inconclusive evidence regarding whether the fetus was still alive once it was removed from the mother's body, or whether Dr. Edelin had caused the death of the fetus by delaying its removal.[89] Despite conflicting testimony and despite the defense that he had done nothing beyond the bounds of standard medical practice, Dr. Edelin was convicted of manslaughter in Massachusetts Superior Court on February 15, 1975.[90]

The verdict sent shock waves through the medical community. According to a survey of Connecticut obstetricians and gynecologists, "reactions by members of the medical community to Dr. Edelin's conviction were swift, and ranged from acerbic outrage to mild approbation." Most of the physicians polled believed that doctors would be reluctant to perform abortions in light of the decision.[91] Twenty-one medical house officers of the New England Deaconess Hospital in Boston signed a letter registering outrage over what they perceived as a threat to medical authority: "It is appalling . . . that in a city that prides itself as representing the best traditions in American medicine, a trial that should be decided on medical expertise is settled instead on histrionics and inflammatory evidence."[92] The editor of the prestigious *New England Journal of Medicine* pronounced the Edelin case "a fiasco" and called for contributions to set up the Kenneth Edelin Defense Fund to support an appeal to a higher court.[93] Certainly the incendiary public nature of the case stemmed, in part, from the fact that the case concerned abortion, an issue that was, and remains, highly controversial. But the case's implications for the potential criminalization of medical practice were not lost on medical professionals. According to Washington, D.C., doctor-lawyer Harold Hirsh, who taught at Georgetown and Catholic univer-sities, "After the Edelin case, doctors are really scared. There's a real concern in the profession over criminal liability."[94]

News reports and analysis did not miss the importance of liability concerns in prompting the Quinlan litigation, although the role of the 1968 redefinition of death in creating this professional distress was not addressed.[95] But the possibility that Karen's doctors had refused to remove her from the respirator due to fears of the criminalization of medical practice was a muted claim, denied by the physicians themselves and lost in the din of more dramatic accounts emphasizing the dangers of a technology out of control.

Erroneous information concerning Karen's situation had been seized upon swiftly and zealously. It inflamed news reports, commentary, and letters from the reading public, and it resisted efforts at correction. The intensity with which these misconceptions were met, harbored, and withstood revision suggests that a cultural animus had, in part, given these renditions of events and meanings their public life. The fear that doctors were creating Frankenstein monsters throughout America was the product of a cultural cast of mind.

The reading public was led to believe that Karen was brain dead and that doctors were keeping her body breathing artificially. "Karen Anne has been in a coma for six months," the *San Francisco Chronicle* told its readers, "her brain dead, her body sustained . . . only by mechanical means." The *Los Angeles Times* reported that the Quinlans believed Karen was dead except for being on the machine. "She's just a vegetable," Mrs. Quinlan was quoted saying, "She is not alive."[96] But, in fact, Karen was alive as judged by any criteria. She was in a "chronic vegetative state," which left her with lower brain stem function, making her responsive to external stimuli like light and sound. Early accounts reported that Karen had detectable brain waves; but it was not made clear that having brain waves is inconsistent with being "irreversibly comatose," or brain dead.

The notion that Karen was brain dead had been suggested early on by the Quinlans' lawyer who, himself, had misunderstood the situation.[97] This claim later was dropped during the trial by a stipulated amendment to the pretrial order. Instead, the court would now be asked to recognize Karen's "right to die."[98] Despite subsequent clarifications, however, the impression that Karen's "dead" body was being artificially maintained never completely disappeared.

Commentary often compounded the mistake by suggesting that legislative adoption of the brain death criteria could help

Karen; if only New Jersey had a brain death statute, the thinking went, then Karen could be pronounced dead and released
from her mechanical hell.[99] Of course, such a statute would not
have helped Karen because all who examined Karen Quinlan
agreed that she was alive by *any* criterion.[100]

Pretrial proceedings began to make the situation clearer. Dr.
Milton Heifetz, a neurologist consulted by the medical experts
on the case, explained that Karen did not fit any definition of
death. "Any attorney who attempts to argue that the woman's
condition would fit a definition of death would be totally in-
correct," Dr. Heifetz told interviewers. "It has no bearing, it
would only confuse the case."[101] Despite news coverage of
such refutation, however, the misunderstanding proved tena-
cious.[102] Ten years after the trial, the *Boston Globe* would still
mistakenly tell its readers that "during a two-week trial in
Morristown, doctors testified that Miss Quinlan had experienced brain death."[103]

The American public also read that Karen's imprisonment
by the respirator was due to unprecedented ethical questions
caused by advancing technology. "The old laws and customs
based on the notion that everyone is either alive or not alive . . .
have been made obsolete by medical progress," the *New York
Times* editorialized.[104] The *New Orleans Times Picayune* observed that when the Quinlan case "broke on an astonished
world" it had been widely supposed that Karen's situation was
a rare, if not unique one.[105] Mechanical ventilation, however,
was a part of medical practice that had been growing steadily
since the 1950s (see Chapter 3).

Readers were also made to feel uneasy about medicine's developing custom regarding how to deal with this technology.
The news coverage reported or implied that doctors were keeping people like Karen ventilated indefinitely. In an op-ed piece
for the *Los Angeles Times*, Georgie Anne Geyer wrote that science was "fearfully defending any form of life at any cost."
Charles Whelan, associate editor of *America* and constitutional
law professor at Fordham Law School, told readers that doctors
had refused to disconnect Karen "on the ground that . . . cessa-

tion of life-sustaining measures was against their conscience and not warranted by accepted medical standards."[106]

In fact, confronting such choices had been a part of medical practice for some time. Wedged between accounts with sensational headlines asserting that "Months in Coma, Young Woman Is a Patient Apart," some reports tried to set news coverage straight. "It is done all the time," New Jersey neurological surgeon Arthur Winter told reporters.[107] Washington, D.C., physician Dr. Michael Halberstam made clear that, "a significant aspect of the Karen Quinlan case is the fact that it is in court at all."[108] As cases similar to Karen's surfaced and journalists became enlightened about actual medical practice, the news registered the fact that physicians allowed patients to die more regularly than previously believed.[109] The belief, however, that physicians "never stop trying to sustain life . . . no matter how grotesque" persisted.[110] It endured, in part, because courtroom evidence supported it.

The Quinlan legal proceedings skewed courtroom evidence presented by physicians. Although some physicians were candid enough to admit to journalists or to publish in remote journals that pulling the plug was standard medical practice, it was quite another matter to admit this in a court of law. The Quinlans' lawyer contacted more than a hundred physicians in an effort to find one who would function as an expert witness for their side of the case. Although many doctors admitted to the "almost 'routine'" practice of removing life support, none would testify to such practice in court. According to the Quinlans, "He found many who were sympathetic, and who confirmed that the removal of life supports in hopeless cases was almost 'routine' hospital practice. But they backed away from stating this truth 'under the white light of litigation.'"[111] Dr. Sidney Diamond, one of the medical experts who testified at the trial, told the lower court that all physicians would have placed Karen on the respirator and that he knew of none who would remove her from it.[112] Based on such testimony, both the superior court and New Jersey's supreme court took customary medical practice—the standard against which the law deter-

mines what is right or wrong in cases involving physicians—
to be that of keeping patients like Karen ventilated.[113]

In a legal symposium held after the Quinlan litigation,
Karen's court-appointed guardian reflected on how the law
came to accept an erroneous view of medical practice. "None
of the other attorneys," he recalled, "would say that [removal
of respirators in situations like Karen's occurs on a daily basis]
at the oral argument. . . . All the doctors told me that this happened all the time, in varying degrees in different situations,
but while no one would come forward and testify to it, it happens. My doctors would have testified to it, but . . . I wasn't going to ask them that." Karen's guardian also related that the
chief justice of the New Jersey Supreme Court said that the
court could *almost* take "judicial notice" of the fact that removal of respirators occurs on a daily basis. Had the court done
so, it would have been in spite of the fact that there was no evidence to support such a conclusion in the record.[114]

The structure of legal proceedings concealed the reality of
a nuanced medical practice—the courts were unable to gauge
the realities of actual medical behavior. The "white light of
litigation" created a false picture of "standard" medical custom and the press relayed this distorted picture to the public.
"None of the witnesses said it would be proper to remove her
from the mechanical respirator that was sustaining her breathing," the public was told, "and one physician said that 'standard medical practice' would dictate continuing the respirator."[115]

Perhaps the most enduring misconception about the case—
that the New Jersey Supreme Court ruling had decreased professional discretion and increased patients' rights—was generated not by the dynamics of litigation per se but by the
specific ruling made by the New Jersey Supreme Court. The
lower court had ruled that Karen should remain on the respirator in accordance with physicians' orders; the higher court
ruled that she could be removed. The response of medical journals to these verdicts is virtually nonexistent. Historian David

Rothman concluded that "many physicians reacted hostilely to the decision, finding *Quinlan* an egregious example of the subversion of their professional discretion."[116] Although some physicians may have expressed modest amounts of concern to newsweeklies at various points along the way before the supreme court's final ruling, there is scant evidence that organized medicine was alarmed by the ruling itself. Of the examples Rothman cites, one is written not by a physician but by a bioethicist, and the second remark is from a physician commenting on matters before either the superior or the supreme court had rendered its decision. There was no tense or sustained discussion in the medical journals themselves fearing any calamity caused by the Quinlan decision; neither was there an effort made on the part of organized medicine to have the Quinlan case overruled either legislatively or judicially. This lack of reaction raises a question: if the implications of the Edelin case and the tense climate created by the medical malpractice crisis had many physicians anxious about the legalities surrounding medical practice, why was the medical profession so apparently unbothered by the New Jersey Supreme Court ruling on the Quinlan case—a ruling that appeared to encroach on professional autonomy?

In fact, neither ruling had much effect on medical practice or custom. Before the lower court ruling, professional medicine did show some modest concern over what kinds of judicial or legislative responses the case might elicit. The American Medical Association (AMA), for example, released a mollifying statement to the public that consent forms specifying familial approval for ceasing extraordinary treatment were routinely required by hospitals. The AMA went on to emphasize that it had developed a policy statement rejecting "mercy killing" but conceding that "removal of extraordinary means to prolong the life of the body when there is irrefutable evidence that biological death is imminent is the decision of the patient and/or their immediate family." As such, the spokesperson for the AMA suggested, no new statutes were required to deal with the matter.[117]

Once the lower court ruled, however, professional medicine saw little cause for alarm. Judge Muir's language seemed to favor medical professional autonomy: "The nature, extent and duration of care by societal standards is the responsibility of a physician. What justification is there to remove it from the control of the medical profession and place it in the hands of the courts?"[118] Medical authorities agreed that the superior court's decision would change very little. At that juncture the important aspect of the case—for the medical profession—was that Judge Muir had viewed the controversy as a medical matter to be left in medical hands. The American Medical Association's president, Dr. Max Parrott, was untroubled by the result: "What Judge Muir said was that the issue was strictly a medical decision. That means, unless restrictive legislation is enacted, doctors will continue to deal with each case on an individual basis. I foresee no change in the way physicians use or discontinue life-sustaining procedures."[119]

The lower court had indicated, however, that removing the respirator *might* constitute homicide under New Jersey's statute. Some "medical ethical experts" predicted that some physicians may grow wary of either "pulling the plug" or even of placing patients on respirators initially.[120] Because the lower court explicitly exonerated the decision-making autonomy of physicians, however, authoritative medical spokespersons appeared unconcerned. "The Quinlan decision [in the lower court]," according to the American Hospital Association, "may make doctors more cautious for a while in decisions on the use of life-sustaining machines, but pretty soon they will be making their decisions on a case-by-case basis as they always have."[121] *U.S. News and World Report* contacted medical authorities from around the world and concluded that the superior court's decision was "unlikely to have any permanent or far-reaching impact on the way medicine is practiced."[122]

In contrast to the lower court's ruling, the New Jersey Supreme Court's much-vaunted acceptance of Karen's "right to privacy" seemed to champion Karen and, in so doing, appeared seriously to challenge medical autonomy. On closer scrutiny, however, the decision reveals itself to be less than a trenchant affirmation of patients' rights in a struggle against medical paternalism. In fact, it gave legal confirmation to a broadly based medical discretion.

Having accepted the claim that removing Karen from the respirator was not an action sanctioned by the medical community, the supreme court's embrace of Karen's right to privacy appeared to contravene medical professional control: "Presumably this right is broad enough to encompass a patient's decision to decline medical treatment under certain circumstances," wrote the court, "in much the same way as it is broad enough to encompass a woman's decision to terminate pregnancy under certain conditions." But what the higher court gave with one hand, it took back with the other. In listing the specifics of the declaratory relief, the court slipped control back into the hands of the medical profession:

> Upon the concurrence of the guardian and family of Karen, should the responsible attending physicians conclude that there is no reasonable possibility of Karen's ever emerging from her present comatose condition to a cognitive, sapient state, *and* that the life-support apparatus now being administered to Karen should be discontinued, they shall consult with the hospital "Ethics Committee" or like body of the institution in which Karen is then hospitalized. If that consultative body agrees that there is no reasonable possibility of Karen's ever emerging from her present comatose condition to a cognitive, sapient state, the present life-support system *may* be withdrawn and said action shall *be without any civil or criminal liability* therefore on the part of any participant.[123]

The court thus stipulated that the physician has the authority not only (1) to determine that there is no reasonable chance that Karen would emerge from her comatose condition to a cognitive, sapient state but also (2) to determine that the life-support system should be discontinued.[124] A great irony of the Quinlan litigation is that the lower court, although it explicitly stressed the importance of preserving a doctor's autonomy, in fact suggested a narrowing of that autonomy by confirming the potential for criminal liability; the higher court, although it explicitly stressed the importance of expanding a patient's right, in fact worked to expand medical autonomy by freeing physicians from criminal liability.

After the New Jersey Supreme Court ruling, professional medicine's response was minimal. The AMA's chief objection was to the court's recommendation that physicians consult with an "ethics committee." Legal counsel for the AMA, B. J. Anderson commented, for example, that such committees were unnecessary, because "a treating physician is certainly able to determine whether a patient is in a terminal condition. If he is unsure of anything, the doctor can ask for consultation with another doctor. Most hospitals don't have 'Ethics Committees.'"[125] This disagreement with one aspect of the court's ruling, however, was over a recommendation, not a requirement, and there was no persistent or widespread censure of the court's decision. Many agreed with the medical director of Georgetown University Hospital, Dr. John Stapleton, that the court's opinion was "realistic" and in "accord with what we would generally practice."[126] The lack of professional perception of calamity stems from the fact that the New Jersey Supreme Court left ultimate discretion with the physician—this despite the higher court's lauded promulgation of the patient's "right to privacy."

Received wisdom about the Quinlan case sees it as a significant shift in the history of medical decision making: "a profession that had once ruled was now being ruled."[127] In fact, the opposite is true. The Quinlan court did not order physicians to turn off the respirator; it merely gave assurance that doctors may do so should they, in their professional wisdom, deem this an appropriate course. The decision was less an offer to Karen of freedom from technology than it was an offer to the medical profession of freedom from liability. American culture in the 1970s, having fostered the chimera that doctors imprisoned the dying bodies of patients, also accommodated the notion that the court had worked readily to free these victims from a vicious medical technology.

Misperceptions surrounding the Quinlan case, fueling fears of technological menace, breathed new life into the nascent bioethics movement. If hospitals across America were filling up with "machine sustained organisms," if modern medical technology posed a threat to the very dignity of human life and death, then surely changes had to be made, doctors had to be watched, medical discretion curtailed. Vigilance and scrutiny were essential, and bioethics stood ready to face the challenge.

"For myself," began one *Los Angeles Times* reader, "I would subject the decision on whether I should live or die, given that I was in Karen's position, to a vote by a committee made up of a philosopher, my spouse or relative, a doctor, a clergyman, and a judge. I know of no single person who is infallible and omniscient."[128] Kenneth Vaux, bioethicist and professor of theology and ethics at the Institute of Religion, Texas Medical Center, in Houston saw matters similarly. Doctors needed help, he urged, in becoming moral decision makers.

> The [Quinlan] case points up again how poorly selected and inadequately trained our physicians are. These are, for the most part, persons with a technical aptitude and a constitutional inhibition against moral analysis and judgment of human situations. Yet the practice of medicine will increasingly require these moral graces. We need to fashion ways to strengthen these capacities in medical training. . . . We cannot continue to "clone" the present generation of doctors. We need to develop courses, clinical experiences, supervised modes of learning that will help a generation of young doctors obtain ethical wisdom equal to their technical competence.[129]

The bioethicists' clarion call had been sounded. Bioethics offered an interdisciplinary "expertise" to protect a society that felt threatened by the overzealous reach of its own medical and technological successes.

The Quinlan case engendered approval of the interdisciplinary, ethics-oriented approaches bioethics offered. The New Jersey Supreme Court had sanctioned interdisciplinary, ethical oversight when it implored hospitals to use "ethics committees" to assist in decisions over whether to discontinue treatment. Testifying to the new found importance of bioethicists, the Quinlans' lawyer consulted at length with both the Kennedy Institute, Center for Bioethics in Washington, D.C., and the Institute of Society, Ethics and Life Sciences (the Hastings Center) in New York to prepare for his appearance before the New Jersey Supreme Court.[130]

Perhaps more important for the long-run development of bioethics as a social institution, ethicists found their opinions in demand; journalists began seeking their commentary for quotable remarks in news analysis.[131] Major newspapers and the newsweeklies, *Time* and *Newsweek*, were the most popular vehicles for bioethical commentary. These contributions, however, were generally limited to the printed equivalent of "sound bites," precluding efforts to clarify misconceptions. In October 1975 *Time* quoted bioethicist Robert Veatch remarking that the only estimate for the number of terminal patients kept alive by artificial means is "lots and lots." A November issue of *Newsweek* provided slightly more space: "There's nothing in medical training that qualifies a physician to make these decisions. If any values count, they should be the patient's and the family's."[132] When called on to comment specifically on the case, bioethicists were quoted drawing modest conclusions or attempting to clarify misunderstandings surrounding brain death or the realities of medical practice. Early bioethical analysis left largely unexplored the political and legal liability issues which had given rise to the case;[133] and lengthier efforts to elucidate the issues appeared in less popular venues.[134]

Although there was some difference of opinion on how the case should have been resolved, bioethicists generally supported Karen's removal from the respirator. At one extreme, Thomas Oden, professor of theology and ethics at Drew University, worried about the slippery slope toward euthanasia and believed that lower court's Judge Muir had been correct in denying the Quinlans their request. "Although the Quinlan family deserves our compassion and sympathy, their anguish does not justify a court sanction ordering the termination of an innocent life in a civil action, or a court order requiring a physician to perform an action which violates his moral conscience and standard medical practice."[135] At the other extreme, retired Harvard ethicist, Joseph Fletcher, believed that practicing "death control" was as important as birth control. He was not disturbed by the slippery slope or the prospect of direct euthanasia itself. "Morally speaking," instructed Fletcher, "there is no real difference . . . between direct and indirect euthanasia. The reason is that the end sought in both forms is the same—to contrive the death of a person in an act of mercy."[136] Later, George Annas, director of the Center for Law and Health Sciences at Boston University, took what proved to be an atypical posture for a bioethicist in that he acknowledged more of the political and legal liability concerns and consequences of the case. Although he did not fault the high court for offering relief to the Quinlans, he did take the court to task for its legal reasoning, which he felt unduly freed physicians from responsibility and accountability.[137]

But the larger theoretical framework adopted by most of those bioethicists who commented on the causes and implications of the case mirrored dominant public concerns and understanding. They challenged "medical suzerainty" over decision making and warned against the evils of uncontrolled technology.[138] The bioethical challenge to medical authority reflected the prevailing discourse of individual rights during the 1970s.[139] The need to control the technological threat to humankind, however, was the defining bioethical issue. The focus was on a technology that had become separated from its social, political, and cultural production and reception. Overlooking the role of the politics of transplant research associated with the 1968 redefinition of death and largely neglecting the importance of removing the threat of criminal liability, bioethical analysis nourished fear and faulty impressions by focusing squarely on the need to come to terms with a preexistent technology inexplicably set loose upon an unprepared world.

Thomas Shannon, then assistant professor of social ethics at

Worcester Polytechnic Institute, believed the Quinlan trial signaled humankind's new yet disturbing relation to technology:

> Much more than the fictional dictatorship of the computer "Hal" in the movie *2001*, the trial of Karen Quinlan may well have revealed something of the real relation between humans and technology. Indeed, the decision that Ms. Quinlan must be artificially maintained by her respirator may be a clear signal that humans stand in a new relation to technology and the products of a technological civilization.
>
> Humanity has spent the last few centuries rejoicing in its freedom from the domination by the powers of nature. Now it may suddenly find its position reversed. Only this time the dependency may be a creature of humanity's own making.[140]

Willard Gaylin echoed this sense of drama and tension. Owing to the advent of new technology, society had produced "artifact-humans":

> The borders between life and death . . . have become obscured by the great successes of modern medicine. . . . We have destroyed the balance that formerly existed whereby death closely followed the loss of those personal qualities that involve life as a human being. We have facilitated the survival of artifact-humans. If we have not created new life forms, we certainly have created a new form of life. We have extended the process of dying into something that approaches a new way of life.[141]

For Kenneth Vaux "sophisticated life-support technologies have ushered in a new era in which we can no longer stand passive and powerless. . . . We cannot develop technological interventions and then disclaim continuing responsibility for their utilization and retraction."[142] Missing from such analyses was any implication of the political choices that helped produce the Quinlan impasse.

In the years following the Quinlan litigation, bioethical texts discussed the case abstractly, without problematizing the nettle of political and liability issues that had given rise to the dilemma. Text considerations of the Quinlan case included viewing the case as a harbinger of the right-to-die movement, as one item in an in seriatim list exemplifying how technological advances necessitate ethical scrutiny, as a way of characterizing the right to refuse treatment, or as a means of elucidating analysis of such philosophical precepts as "the principle of non-malfeasance" in ethical decision making.[143]

When news of the Quinlan tragedy first hit the presses, published responses were divided between those wanting to protect society's weakest from the slippery slope toward euthanasia, and those believing that Karen and her family should not have to endure ghastly and prolonged suffering. Eventually, the sensibility that Karen's grotesque imprisonment by the machine was an indignity that none should have to bear came to dominate public discourse. A cultural predisposition that feared swollen medical authority and perilous technology fueled, and was fueled by, misconceptions surrounding the Quinlan litigation. In particular, misunderstandings regarding brain death, the respirator, and the customary behavior of medical practice obscured the legal and political factors that erupted into litigation.

Anxieties over the social threat seemingly implicated by the Quinlan impasse encouraged the establishment and growth of bioethics as a social institution. Bioethical analysis mirrored the concerns of society at large, focusing on curtailing medical discretion and controlling technology. When called on for journalistic commentary, some ethicists clarified the sources and nature of the Quinlan case in printed "sound bites." Serious bioethical discourse, however, limited analysis by overlooking the long-term political genesis and reception of medical technology. The fact that *In Re Quinlan* was more about freeing physicians from liability than about establishing patients' rights was mainly ignored.

Karen's tragedy now informs abstract discussions in ethics texts; nevertheless, it is the forgotten political ramifications of the case that underpin the role in history that she predicted for herself the summer of 1974. These ramifications are put into bold relief by her story after the litigation. Karen's physicians did not disconnect her body from the machine. As they had done before, her physicians tried to "wean" Karen off the ventilator in an effort to get her to breathe on her own—a strategy that ultimately worked. The physician who had functioned as the Quinlans' expert witness told Karen's father, "You have to understand all the newspapers in the country are looking over his [the doctor's] shoulder right now. Don't push him." Almost six weeks after the New Jersey Supreme Court's ruling, Karen was not only still attached to the ventilator, but more medical technology was being employed, as her physician ordered a body-temperature control machine in an effort to curb a fever.[144] These exertions were ultimately effective; weaned off the respirator, Karen "lived" another nine years.[145] *In Re Quinlan* had not required otherwise.

Conclusion and Outlook

"Unless specific counter measures are taken," warned futurist Alvin Toffler in 1970, "if something *can* be done, someone, somewhere *will* do it. The nature of what can and will be done exceeds anything that man is as yet psychologically or morally prepared to live with." "Technology," he urged, "cannot be permitted to rampage through the society." The process had to be controlled, Toffler believed, because people already had begun suffering from what he called the disease of change, future shock, which he defined as "the shattering stress and disorientation that we induce in individuals by subjecting them to too much change in too short a time." For Toffler, the *rate* of change had effects on people that were, at times, more important than the *directions* of change.[1] Like Toffler's widely read book *Future Shock*, bioethics grew out of the sixties to minister to a society suffering from the disease of change. Bioethics delivered exotic technologies into broad social acceptance.[2]

But although bioethics developed in the 1960s, it was not simply of the 1960s. The initial bioethical impulse—to find certain technological advances disturbing—is part of a recurrent pattern in American history. Specific historical circumstances, however, did give rise to bioethics. The anxieties of geneticists working in the wake of the postwar responsible science movement were an early part of the social forces pushing bioethics into existence. Yet, even while distress over biotechnology was erupting, the process of dissipating that distress also had begun. The Hastings Center offers an intimate example of how established medical and scientific interests constricted potentially threatening avenues of inquiry and quelled the more challenging instincts of early bioethicists seeking institutional security. The decade's hallmark attitude toward technology, fraught with alarm, abated as bioethics began the process of institutionalization.

The redefinition of death and the Quinlan case lend insight into the simultaneous institutionalization of bioethics and diffusion of fear over biotechnology. Professional demand for bioethical services grew as supporters of transplantation research harnessed bioethical energies to assist in redefining death in the effort to manage public opinion. The case of Karen Quinlan shows the human consequences of an alliance between medical research interests and medical research oversight of this redefinition. Moreover, the response to misconceptions surrounding the Quinlan case accentuated the public's demand for bioethical services.

Previous generations expressed their ambivalence toward technological development in ways unique to their historic culture; they may have worried, for example, over the sullying of the republic or perhaps the decline of civilization. The cultural idiom of the sixties consisted of broad attacks against the authority and power of established social institutions. The bioethics movement was not a simple outgrowth of the critical forces endemic to the sixties. It was not a radical movement. It did, however, tilt along the axis of challenge that was the decade's hallmark. Established research interests and funding agencies constrained the impetus of challenge that early bioethics possessed. In this subdued incarnation, bioethics served to transmute potentially hostile impulses of the larger society into an acceptable expertise. Bioethics did not become an established source of critical oversight. It became, instead, an opportunity for ethics experts to cope with dilemmas generated by technologies, which came to be seen, ironically, as value-neutral in their creation even while they were problem-causing in their effects.

Considered within the larger narrative of the 1960s protest culture, the role of bioethics from the 1960s to the 1970s may be seen as one in the transition from critique to management. The cultural politics of bioethics is a history of the waning of the sixties milieu of challenge. It is the story of the dissipation of one generation's anxieties about technology—the recurring anxieties of a technological nation.

The transmutation of cultural hostility into professional expertise, a legacy of bioethics, has had consequences that affect us today. Examining continuing ramifications of the Quinlan case, for example, makes this clear. March 1997 marked the twentieth anniversary of *In Re Quinlan*. Karen's place in history was remembered in newspapers, radio shows, and conferences across the country as having launched the right-to-die movement. The central feature of the larger social movement is the body of legal cases expanding the patient's "right" to decline life-sustaining treatment, to compel medical professionals to withdraw such treatment if administered, and ultimately, perhaps, to require professional assistance for terminally ill individuals wishing to end unendurable lives.[3] As we have seen, however, this is a false genealogy. Stripped of its drama, the Quinlan case does not mark the beginning of the movement to deal compassionately with terminally ill patients. It marks the beginning of this movement's fatal turn—from the ambit of patients, doctors, families, clerics, and communities to the sphere of the courtroom and the language and limitations of "rights."

This turn resulted from the combination of medical research aims and legal liability fears surrounding organ transplantation, considerations that had little to do with humane concern for the irreversibly comatose or the terminally ill (see Chapter 4). If the Quinlan case can be remembered for these aims and fears, it can help to explain two continuing controversies regarding views about death: the ongoing disagreement over the definition of death and the public debate over physician-assisted suicide.

It is not surprising that the 1968 definition of brain death wreaked havoc for the Quinlans in the 1970s. The definition was adopted, after all, not because of an exigent need to deal with the nature of death on its own terms but over the interest in facilitating organ procurement. In a more academic way, the definition continues to cause problems. Difficulties persist over the desire for a logical definition of death considered on its own terms on the one hand, and a definition that will allow for the harvesting of transplantable organs on the other. The traditional cardiorespiratory definition, which most find acceptable on its own terms, has the disadvantage of leaving no room for the removal of transplantable organs. As discussed in Chapter 3, waiting for death under this definition renders the organs unusable. On the other hand, while the brain death definition (subsequently formulated as the "whole brain" concept of death) raises the probability of yielding usable organs, it also has kept practitioners and ethicists debating because it is logically inconsistent and riddled with technological hazards. The "whole brain"—brain death definition, although accepted broadly, is not accepted without argument.

Harvard Medical School associate professor of anesthesia and pediatrics Robert Truog offered a succinct analysis of the controversy in a 1997 article.[4] "There is evidence," he writes, "that many individuals who fulfill all of the tests for brain death do not have the 'permanent cessation of the entire brain.'" "In particular," he continues, "many of these individuals retain clear evidence of integrated brain function at the level of the brainstem and mid-brain, and may have evidence of cortical function." Truog notes that the technology that can demonstrate adequately "whole brain" destruction does not exist; this can be determined only at autopsy. By the time of autopsy, of course, it is too late to recover usable organs.

Truog goes on to assess the "higher brain" criterion of death, proposed in the mid-1970s (after Quinlan) as an alternative to the brain death criteria.[5] Proponents of this criterion focus on consciousness as the critical function of the brain and conclude that all those who are permanently unconscious can be declared dead. This standard would include those who have too much brain function to be considered dead by the "whole brain" criterion. (Karen Quinlan could have been declared dead by this criterion.) Here, again, Truog asserts that the capacity for diagnostic certainty is overstated. Truog also outlines problems with a definition of death that focuses on the death of the *person* rather than of the *organism,* and which leaves room for treating breathing people as if they were dead. He also notes correctly that "at the present time any inclination toward a higher brain death standard remains primarily in the realm of philosophers and not policymakers."[6]

The way to settle problems over diagnostic inadequacies and logical inconsistencies, according to Truog, is to have two separate criteria: one for establishing death, and one for designating potential organ donors. To establish when death has occurred, we should, he urges, return to the traditional cardiorespiratory standard for death. To designate potential organ donors, Truog offers directions, not specific criteria. Policies could be formulated to insure that procuring organs happens only "with the consent of the donor or appropriate surrogate and only when doing so would not harm the donor." Under Truog's approach, "individuals who had given their consent could simply have their organs removed under general anesthesia, without first undergoing an orchestrated withdrawal of life support." He makes plain that with this approach, organ procurement would have to be seen as a "form of justified killing," and not just "the dissection of a corpse." According to one of Truog's critics, "This initially attractive idea produces serious practical problems because essentially it certifies physicians to kill dying patients to procure their organs."[7]

Truog's proposal underscores the role of bioethics as broker of exotic biotechnologies. In 1968, when researchers fretted that the new need for finding transplantable organs would alarm a public unaccustomed to viewing organs as fungible, the concept of brain death was a necessary balm, sanctioned by bioethicists. Today, as Truog suggests, the concept may be obsolete.

The diagnosis of brain death has been extremely useful during the last several decades, as society has struggled with a myriad of issues that were never encountered before the era of mechanical ventilation and organ transplantation. As society emerges from this transitional period, and as many of these issues are more clearly understood as questions that are inherently unrelated to the distinction between life and death, then the concept of brain death may no longer be useful or relevant.[8]

Currently, artificially ventilated patients who are not considered potential organ donors are not necessarily diagnosed as brain dead; often, artificial ventilation is simply withdrawn.[9] Declaring a patient brain dead before withdrawing life support, according to Truog, is no longer considered crucial unless seeking organs. It remains to be seen whether the medical profession as a whole, much less the public, would be willing to accept organ donation without first requiring a declaration of death. Truog, nevertheless, seems correct in signaling a degree of comfort with the idea of organ transplantation that was not evident in the late sixties when the brain death criteria were institutionalized.

Truog's delineation and the attacks against it are the most recent volleys in a decades-old debate that only occasionally draws popular interest. For the most part, the discussion remains in the domain of ethicists and practitioners.[10] An area in which the public is involved deeply, however, is physician-assisted suicide. Here, too, remembering the Quinlan case rightly can help explain the nature of this current social impetus and its legal expression.

Viewing the Quinlan case incorrectly as having single-handedly launched the right-to-die movement ignores the context of public and professional expression of anxieties over the need to curtail modern medicine's prolongation of suffering, which dates back to the 1950s. It also overlooks how during the 1970s (set up by the deliberately confounding genesis of the 1968 redefinition of death) early press coverage of the Quinlan saga encouraged the mistaken impression that doctors had been insisting that chronically vegetative people who were being artificially ventilated should not be removed from ventilation. In this context, what other recourse did members of the public have but to invoke "a right to die" to protect against such horrifying abuse by medical practice? Remembered wrongly, the Quinlan case's logical conclusion twenty years later may be Kevorkian—the maverick "doctor of death," impassively arriving to administer death to people scarcely known to him a short time before—who, grotesque as it may be, has come to represent the patient's final right.

The U.S. Supreme Court has not signaled a willingness to support this conclusion. In 1990 the Court declined the possibility of taking the Quinlan case a decisive step further when it considered the case of *Cruzan vs. Missouri Department of Health.* Rescuers found Nancy Cruzan too late to restore her to full consciousness after she lost control of her car on a Missouri road in 1983. Like Karen before her, Nancy now existed in a persistent vegetative state. In time, Nancy's parents came to request that physicians remove the gastronomy tube that surgeons had implanted to feed and hydrate their daughter. The Supreme Court did not support their request. The Court assumed that "a competent person would have a constitutionally protected right to refuse lifesaving hydration and nutrition." But with respect to an incompetent person, the Court reasoned that the state's interest in preserving life legitimated a requirement that clear and convincing evidence be offered to demonstrate the wishes of the incompetent adult patient expressed before that patient had been rendered incompetent. Ruling that this standard had not been met in Nancy Cruzan's case, the Court denied the request to remove the tube.[11]

Both the Quinlan and Cruzan cases are part of the right-to-die movement in that they concern a right to refuse medical treatment, something that would, it was believed, lead to death. Eventually, the U.S. Supreme Court considered the ultimate right to die: the right to receive medical assistance to end one's life deliberately. In 1997 the Court decided unanimously to decline recognition of a general, constitutionally protected right to commit suicide with physician assistance (by means of lethal injection). In *Washington vs. Glucksberg* the Court upheld a Washington State statute making it illegal to cause or assist in a suicide, and in *Vacco vs. Quill* the Court upheld a similar New York State statute.[12] Both cases involved physicians who were willing to assist dying patients to commit suicide. While the justices found no general constitutional right to assisted suicide, they left some room for reconsidering the issue at a later date.[13]

If we remember the Quinlan case wrongly as triggering the right-to-die movement, then the cases involving Cruzan, Glucksberg, and Vacco may be seen as ending it. But the movement is older and larger than its judicial expression. In deciding *Washington vs. Glucksberg*, the Court seemed to recognize this, at least in part, when it acknowledged that the Court should not have the last word: "Throughout the Nation, Americans are engaged in an earnest and profound debate about the morality, legality, and practicality of physician-assisted suicide. Our holding permits this debate to continue, as it should in a democratic society."[14] While the Court upheld state laws that prohibited assisted suicide, it implied that there is not necessarily a constitutional restriction against a state statute that might specifically allow it. Dozens of states have passed laws prohibiting assisted suicide. Michigan voters rejected a proposal for a law permitting physician-assisted suicide in their November 1998 elections. The state of Oregon, however, has supported physician-assisted suicide.[15]

The Quinlan case has taken us down a path where each state may attempt to choose for itself whether to grant the ultimate "right" to die, assisted suicide. Remembered wrongly, the case has served to constrain the national discourse surrounding our expectations for humane treatment during chronic or terminal illness; for this discourse now is contorted by a jurisprudence wedded to the strange oxymoronic slogan that we have a right to die—suggesting improbably that, somehow, we have the right to do something that we cannot escape. In a supreme act of denial, we have enlisted the courts and legislatures to help us valorize the disease-compromised prerogatives of the dying instead of developing the obligations of the healthy—the compassionate obligation to seek ways to ease and end suffering. Historically, as we have seen, this exploration of empathy began with our families, friends, physicians, churches, and communities. There it should remain.

There is, in the United States, a strong cultural force supporting the individual's desire to seek swift, painless, dignified death. The signs of this force are clear: right-to-die litigation and legislation; the fact that juries tend not to not find physicians guilty for assisting suicides; the creation of living wills; the existence of organizations like the Hemlock Society, the Euthanasia Research and Guidance Organization, Choice in Dying, and Americans for Death with Dignity; the success of the "how-to" book *Final Exit;* and, until recently, the grim triumphs of the Doctor of Death's sterile administrations.[16] The fact that Kevorkian repeatedly escaped prosecution (until his flagrant dare on the *60 Minutes* episode that broadcast him euthanizing a patient) only further testifies to the existence of strong support for assistance in dying. During the 1970s, the general arrogance and paternalism of the medical profession compelled many to believe that the best way of ensuring death with dignity was to assert a legal "right" to die. The time has come to assess the wisdom of this choice. Bioethicist Arthur Caplan is correct to question the humanity of a jurisprudence that would find a right to die before finding a right to health care.

One of the fears commonly expressed by opponents of physician-assisted suicide is that slowly, but inevitably, the "right" to die will be felt by many as a duty.[17] It remains to be seen whether such fears will be realized in practice. But the groundwork already is being laid. In their celebration of the right-to-die movement, Derek Humphry and Mary Clement urge the merits of what they call "the unspoken argument" that supports physician-assisted suicide: the rising cost of health care. Where better to save dollars than to allow the dying and the elderly to voluntarily end their lives before large sums are spent on them? "A rational argument can be made," exhort the authors of *Freedom to Die*, "for allowing [physician-assisted suicide] in order to offset the amount society and family spend on the ill, *as long as it is the voluntary wish* of the mentally competent terminally and incurably ill adult."[18] Yet the authors do not sufficiently attend to, and seem little bothered by, the problem of how to assure that the decision remains voluntary.

Moreover, philosophical argument has been put forth to the effect that there is, in fact, a duty for individuals to end their lives before the strains and costs of caring become too burdensome for families and society. In a 1997 *Hastings Center Report*, ethicist John Hardwig argued that "there are circumstances when we may have a duty to die."

> As modern medicine continues to save more of us from acute illness, it also delivers more of us over to chronic illnesses, allowing us to survive far longer than we can take care of ourselves. It may be that our technological sophistication coupled with a commitment to our loved ones generates a fairly widespread duty to die.[19]

The tenor of the debate may be shifting from compassion to cost-consciousness. Indeed, in the United States, where consideration of costs figures so largely in the creation of medical standards of care, and where the legal expression of the movement to ease dying was launched not through a consideration of humane treatment but through a response to the instrumental preoccupations of research aims and liability fears, one might begin to wonder how long it will be before our final right becomes our last duty.[20]

The role of bioethics in shepherding the 1968 brain death criteria into acceptance left the way open for creating the false genealogy of the Quinlan narrative and its aftermath, developments that have served to impoverish the discussion over how to care for the nation's dying. More generally, as long as bioethics remains committed to the limited role of establishing guidelines for the use of procedures and technologies that it largely accepts as inevitable, it precludes itself from seeking answers to and informing the public about how and why specific biomedical technologies were created: by whom, or by what groups; on what criteria; for the benefit of which group (race, class, or gender) and to whose detriment; how were such technologies tested; at what cost were they developed, both socially and economically, and instead of what alternatives? Upon what basis and how often does bioethics, as an institutional force, ever recommend prohibiting (as opposed to delaying) the practice or further development of any new biomedical procedure? Arguably, the predictable process of recommending delay until guidelines can be developed has worked to stifle vigorous examination of issues that can be generated most successfully when prohibition is a believable threat. Bioethics may ultimately be successful in helping to alleviate national anxieties about the right to die or in midwifing developments as disturbing as the cloning of a human being. But will it be able to free itself from the sources that help generate the dilemmas it seeks to resolve?

Notes

Preface

1. Adapted from "Moral Specialist," *Newsweek*, January 8, 1979, 6.

2. Elaine Freeman, "The God Committee," *New York Times Magazine*, May 21, 1972, 30–32.

3. Alan L. Otten, "Ethics Experts Help More Doctors Handle Hard Moral Decisions," *Wall Street Journal*, March 6, 1987, 1.

4. Among these institutions are the Hastings Center, Institute of Society, Ethics and the Life Sciences in New York; the Center for Bioethics of the Kennedy Institute in Washington, D.C.; the Midwest Bioethics Center of Kansas City, Missouri; the Bioethics Consultation Group of Berkeley, California; the U.C.S.F. Institute for Health Policy Studies in San Francisco; the Bioethics Council of the Pacific Institute in Solvang, California; and the Center of Bioethics, St. Joseph Health System in Orange, California.

5. See, for example, Ivan Illich, *Medical Nemesis: The Expropriation of Health* (New York: Pantheon Books, 1976), and Theodore Roszak, *The Making of a Counter Culture: Reflections on the Technocratic Society and Its Youthful Opposition* (New York: Anchor Books, Doubleday, 1969).

6. The analysis offered here is not exhaustive. The chapter's early focus is on critiques of technological progress from within the liberal tradition. It provides a context for understanding bioethics' caution and ambivalence regarding the claims of progress—but not criticism of the overall political system.

Prologue

1. I am discussing American ambivalence toward progress and technology. For a more generalized tendency toward romanticized

nostalgia, see Raymond Williams, *The Country and the City* (New York: Oxford University Press, 1973). See also Langdon Winner, "On Criticizing Technology," in *Technology and Man's Future*, ed. Albert H. Teich, 2d ed. (New York: St. Martin's Press, 1977), 354–75. Winner discusses "the radical critics" (of technology) and contrasts them with the "more orthodox scholarship." He maintains that at the heart of the controversy between these two groups "lies an idea which has plagued European and American authors for the last two centuries—the notion that technology has gotten out of control" (360).

2. See T. J. Jackson Lears, *No Place of Grace: Antimodernism and the Transformation of American Culture, 1880–1920* (New York: Pantheon Books, 1981), 26. See also Leo Marx, *The Machine in the Garden: Technology and the Pastoral Ideal in America* (1964; Oxford: Oxford University Press, 1981).

3. See John F. Kasson, *Civilizing the Machine: Technology and Republican Values in America, 1776–1900* (New York: Penguin Books, 1977).

4. Thomas Jefferson, *Notes on the State of Virginia* (London, 1787), 165.

5. Kasson, *Civilizing the Machine*, 38 (quoting Benjamin Franklin).

6. See Charles Sellers, *The Market Revolution: Jacksonian America, 1815–1846* (New York: Oxford University Press, 1991), 44.

7. Henry David Thoreau, *Walden* (1854; New York: Bantam Books, 1982), 116, 124, 135, 193. For other transcendentalists who shared Thoreau's ambivalence, see Roderick Nash, *Wilderness and the American Mind*, 3d ed. (New Haven: Yale University Press, 1982), 94.

8. Thomas P. Hughes, *American Genesis: A Century of Invention and Technological Enthusiasm, 1870–1970* (New York: Viking Penguin, 1989), 298, 1.

9. Lears, *No Place of Grace*, 57–58. Lears offers another observation that also captures something of the social ramifications of bioethics: "Puritan and republican jeremiads have often served to reinforce the dominant culture by reducing social conflicts to questions of individual morality" (6).

10. See Carroll Pursell, *The Machine in America: A Social History of Technology* (Baltimore: Johns Hopkins University Press, 1995).

11. Hughes, *American Genesis*, 9.

12. Pursell, *The Machine in America*, 230.

13. Charles A. Beard, *Whither Mankind: A Panorama of Modern Civilization* (1928; Freeport, N.Y.: Books for Libraries Press, 1971), 1.

14. See especially Bertrand Russell's essay, "Science," in Beard, *Whither Mankind,* 63–82, 5, 14.

15. Beard, *Whither Mankind,* 14, 24, 404.

16. Stuart Chase, *Men and Machines* (New York: Macmillan, 1929), 330. See also Friedrich Georg Juenger, *The Failure of Technology: Perfection without Purpose* (Hinsdale, Ill.: Henry Regnery, 1949).

17. Chase, *Men and Machines,* 347.

Chapter One: The Culture of Post-atomic Ambivalence

1. Quoted in David Rothman, *Strangers at the Bedside: A History of How Law and Bioethics Transformed Medical Decision Making* (New York: Basic Books, 1991), 209.

2. Oral History of Daniel Callahan by M. L. Tina Stevens, Hastings Center, Briarcliff Manor, N.Y., July 1, 1991, transcript, 10–12.

3. Charles A. Beard, *Whither Mankind: A Panorama of Modern Civilization* (1928; Freeport, N.Y.: Books for Libraries Press, 1971), 19.

4. On American culture and the H-bomb, nuclear testing and the arms race, see William O'Neill, *American High: The Years of Confidence, 1945–1960* (New York: Free Press, 1986).

5. Paul Boyer, *By the Bomb's Early Light: American Thought and Culture at the Dawn of the Atomic Age* (New York: Pantheon Books, 1985), 268.

6. O'Neill, *American High;* Boyer, *By the Bomb's Early Light,* 274.

7. Boyer, *By the Bomb's Early Light,* 271.

8. Cited in ibid., 274.

9. Ibid., 49. Boyer believes that the activism of the atomic scientists was, for the most part, a historically unique development: "One can link this activism to a longer tradition of scientific involvement in public issues. (In the activist 1930s, for instance, a few scientists . . . formed the American Association for Scientific Workers and issues pronouncements on current social issues.) But the most striking feature of the postwar scientists' movement was its sudden and spontaneous emergence" (50).

10. Morton Grodzins and Eugene Rabinowitch, eds., *The Atomic Age: Scientists in National and World Affairs* (New York: Basic Books, 1963), v, vii.

11. Boyer, *By the Bomb's Early Light,* 61.

12. Jean Rostand, *Can Man Be Modified?* trans. Jonathan Griffin (New York: Basic Books, 1959), 21. See also Rostand, "Biology and the Burden of Our Times," *Bulletin of the Atomic Scientists* 8, no. 6 (August 1952): 176–78.

13. Gordon Wolstenholme, ed., *Man and His Future: A Ciba Foundation Volume* (Boston: Little, Brown, 1963), 289. But see p. 285 where Bronowski challenges this point of view.

14. See, for example, Hermann J. Muller, *Studies in Genetics* (Bloomington: Indiana University Press, 1962); Joshua Lederberg, *A View of Genetics* (Stockholm: Les Prix Nobel, 1958); Julian Huxley, *The Humanist Frame* (London: Allen and Unwin, 1961); B. Medawar, *The Future of Man* (London: Methuen, 1958).

15. For a history of the Ciba Foundation, see F. Peter Woodford, *The Ciba Foundation: An Analytic History, 1949–1974* (New York: Associated Scientific Publishers, 1974).

16. Gordon Wolstenholme, preface to Wolstenholme, *Man and His Future*, v.

17. Ibid., 397.

18. Hermann J. Muller, "Genetic Progress by Voluntarily Conducted Germinal Choice," in Wolstenholme, *Man and His Future*, 252, 254, 257, 258.

19. Joshua Lederberg, "Biological Future of Man," in Wolstenholme, *Man and His Future*, 264, 265, 266, 268.

20. Wolstenholme, *Man and His Future*, 275.

21. Ibid., 294.

22. Salvador E. Luria, "Molecular Genetics—A Key to Biological Progress," in *The Control of Human Heredity and Evolution*, ed. Tracy Morton Sonneborn (New York: Macmillan), 2–4.

23. Guido Pontecorvo, "Prospects for Genetic Analysis of Man," in Sonneborn, *The Control of Human Heredity and Evolution*, 82.

24. Dr. Rollin D. Hotchkiss, in Sonneborn, *The Control of Human Heredity and Evolution*, 38.

25. Sonneborn, *Human Heredity and Evolution*, vii.

26. Pontecorvo, "Prospects for Genetic Analysis of Man," 81, 82.

27. Concise summaries of these critics' viewpoints may be found in Bruce O. Watkins and Roy Meador, *Technology and Human Values: Collision and Solution* (Ann Arbor: Ann Arbor Science Publishers, 1978).

28. Lewis Mumford, *Technics and Civilization* (New York: Harcourt, Brace, 1934). See also Thomas P. Hughes, *American Genesis:*

A Century of Invention and Technological Enthusiasm, 1870–1970 (New York: Viking, 1989), for his discussion of Mumford.

29. Hughes, *American Genesis*, 448.

30. Mumford, *The Myth of the Machine: The Pentagon of Power* (New York: Harcourt, Brace, Jovanovich, 1970), frontispiece.

31. Ibid., 240.

32. Ibid., 255.

33. Herbert Marcuse, *One Dimensional Man: Studies in the Ideology of Advanced Industrial Society* (Boston: Beacon Press, 1964), ix.

34. Ibid., 32.

35. Mumford, *The Myth of the Machine*, 433.

36. Jacques Ellul, *The Technological Society* (1964; New York: Alfred A. Knopf, 1970), xxvii, 418.

37. Theodore Roszak, *The Making of a Counter Culture: Reflections on the Technocratic Society and Its Youthful Opposition* (New York: Anchor Books, Doubleday, 1969), 5, 8, 9.

38. Ibid., 51.

39. Leroy G. Augenstein, *Come Let Us Play God* (New York: Harper and Row, 1969), 3, 12.

40. See, for example, Gordon Rattray Taylor, *The Biological Time Bomb* (New York: World Publishing, 1968); Gerald Feinberg, *The Prometheus Project: Mankind's Search for Long-Range Goals* (Garden City, N.Y.: Doubleday, 1968); Amitai Etzioni, "Sex Control, Science, and Society," *Science* 161 (September 13, 1968): 1107–12; Albert Rosenfeld, *The Second Genesis: The Coming Control of Life* (Englewood Cliffs, N.J.: Prentice-Hall, 1969); Donald Fleming, "On Living in a Biological Revolution," *Atlantic Monthly* 223 (February 1969): 64; Augenstein, *Come Let Us Play God;* Paul Ramsey, *Fabricated Man: The Ethics of Genetic Control* (New Haven: Yale University Press, 1970).

41. Taylor, *The Biological Time Bomb*, 11.

42. Frances C. Locher, ed., *Contemporary Authors*, vol. 105 (Detroit: Gale Research, 1982), 487.

43. Taylor, *The Biological Time Bomb*, 219–20.

44. Etzioni, "Sex Control, Science, and Society," 1110.

45. Donald Fleming, "On Living in a Biological Revolution," *Atlantic Monthly* 223 (February 1969): 68, 70.

46. For a presentation of the philosophical systems and ethical norms that characterize bioethical literature, see the work of Renee Fox referenced in Chapter 2 and Paul T. Durbin, *A Guide to the Cul-*

ture of Science, Technology, and Medicine (New York: Free Press, a division of Macmillan, 1984).

47. Deontological principles are those whose rightness is not determined by their consequences. Teleological principles are those whose rightness is determined by their consequences. See, for example, H. Tristram Engelhardt, *The Foundations of Bioethics* (New York: Oxford University Press, 1986), 82.

48. Daniel Callahan, "Values, Facts and Decision-Making," *Hastings Center Report* 1, no. 1 (1971): 1.

49. Walter F. Mondale, "The Issues before Us," *Hastings Center Report* 1, no. 1 (1971): [4–5]. It is, of course, arguable whether Mondale should be invoked as a bioethicist; at the time of these statements, however, he was a fellow of the Hastings Center. He made these remarks by way of introducing on the Senate floor a bill that would begin a National Advisory Commission on Health, Science, and Society.

50. Robert M. Veatch, *Case Studies in Medical Ethics* (Cambridge, Mass.: Harvard University Press, 1977), v.

51. George H. Kieffer, *Bioethics: A Textbook of Issues* (Reading: Addison-Wesley, 1979), iv, xv.

52. H. Tristram Engelhardt Jr., *The Foundations of Bioethics* (New York: Oxford University Press, 1986), 4. Other examples include "the issues of bioethics have captured the contemporary mind because they represent major conflicts in the area of technology and basic human values, those dealing with life, death, and health," in *The Encyclopedia of Bioethics* (New York: Free Press, 1978), xv; and "Bioethics . . . includes the questions raised by new biological research," in Ann E. Weiss, *Bioethics: Dilemmas in Modern Medicine* (Hillside, N.J.: Enslow Publishers, 1985), 3.

53. Such reasons include a growing openness to interdisciplinary work; a manifestation of the "explosion of knowledge" characteristic of "our era" (Warren T. Reich, "Introduction," in *Encyclopedia of Bioethics*, xv); rising health care costs which evoke questions regarding the allocation of resources; the pluralistic context in which health care is delivered, in which people no longer share a common normative view; the expansion of rights to self-determination (Engelhardt, *The Foundations of Bioethics*, 4–5); the concentration of medical care in hospitals, the growth of scientific medicine, and the specialization of physicians into narrow fields of study and practice, all of which contributed to an impersonal, organizational approach to health care; the rise of mass media, which popularized new di-

rections in medicine; increasing federal support of biomedical research during the 1950s and 1960s which served to focus public attention on such research; increasing consumerism and its attendant consumer advocate posture (Albert R. Jonsen, Andres L. Jameton, and Abbyann Lynch, "History of Medical Ethics: North America in the Twentieth Century," in *The Encyclopedia of Bioethics*, 992–1004).

54. Arthur L. Caplan, "What Bioethics Brought to the Public," in "The Birth of Bioethics," ed. Albert R. Jonsen, special suppl., *Hastings Center Report* 23, no. 6 (1993): S14-S15. See also the early and later thoughts of bioethicist Robert Veatch. In his original edition, *Death, Dying, and the Biological Revolution* (New Haven: Yale University Press, 1976), Veatch focused singularly on the "technological and biological revolutions." "The biological revolution," he wrote, "has challenged us in our dying and our death, much as it has challenged us in our living. . . . In a simpler day we often knew, or thought we knew, what our objectives were for medicine. . . . The biological revolution poses for us new problems" (vii-viii.)

This same concentration on technology dominated the 1989 revised edition of this work. However, at the Birth of Bioethics Conference, Seattle, Washington, September 23–24, 1992, Veatch went further and identified two "major social developments" in accounting for the development of bioethics (with respect to death and dying issues): "First, was a technological revolution. . . . But second, and I suspect more important, there was a social revolution—the civil rights movement of the 1960s, the anti–Vietnam War movement, the student activist movement, and the women's movement." Robert M. Veatch, "From Forgoing Life Support to Aid-in-Dying," in "The Birth of Bioethics," ed. Albert R. Jonsen, special suppl., *Hastings Center Report* 23, no. 6 (1993): S7.

55. In his effort in 1993 to pinpoint the "birth of bioethics," for example, philosopher and bioethicist Albert Jonsen reflected on the historical precedent of the 1945 Nuremberg Trials over Nazi experimentation. He concluded, however, that "horrible as [Nazi abuses] were, the problem posed by medical research was deeper. It was not the maliciousness and callousness of scientists but the very nature of modern biomedical science that created the problem." Thus, rather than viewing Nazi abuses as significant triggers leading to the development of bioethics, Jonsen was more inclined to target events like the development of kidney dialysis in 1961 or the "dramatic medical advance" of heart transplantation in 1967, or "the problems

posed by the new life-support technologies." Albert Jonsen, "The Birth of Bioethics," *Hastings Center Report* 23, no. 6 (1993): S1-S4. Later, in his 1998 history of bioethics, Jonsen explained that he and his colleagues "knew that disciplines are not born; they grow slowly." Nevertheless, despite a discussion that includes other developments, his narrative is weighted by its emphasis on technologies, procedures or experiments that occur in or after the 1960s.

56. Joseph Fletcher, *Morals and Medicine* (Princeton, N.J.: Princeton University Press, 1954).

57. Daniel Callahan, "The Development of Biomedical Ethics in the United States," in Daniel Callahan and G. R. Dunstan, eds., *Biomedical Ethics: An Anglo American Dialogue* (New York: New York Academy of Sciences, 1988), 1–3.

58. See also "Three Views of History," in "The Birth of Bioethics," ed. Albert R. Jonsen, special suppl., *Hastings Center Report* 23, no. 6 (1993): David Rothman, "View the First," S11-S12 ; Daniel M. Fox, "View the Second," S11-S13; Stanley J. Reiser, "View the Third," S13-S14. Also see in the same issue, Warren T. Reich, "How Bioethics Got Its Name," S6.

59. Albert R. Jonsen, *The Birth of Bioethics* (New York: Oxford University Press, 1998), xiii, 134.

60. Ibid., 137; Rothman and Katz, cited in ibid.

61. Birth of Bioethics Conference.

62. Renee C. Fox, "The Sociology of Bioethics," in *The Sociology of Medicine: A Participant Observer's View*, ed. Alex Inkeles (Englewood Cliffs, N.J.: Prentice-Hall, 1989), 224–76, at 224, 225. See Fox, p. 225, n. 4, which lists other relevant articles by the author.

63. Ibid., 226.

64. Ibid., 230, 231.

65. Rothman, *Strangers at the Bedside*, 243, 245.

66. Ibid., 246.

67. See Rothman, "View the First," S11.

68. David Rothman, "Human Experimentation and the Origins of Bioethics in the United States," in *Social Science Perspectives on Medical Ethics*, ed. George Weisz (Dordrecht, The Netherlands: Kluwer Academic Publishers, 1990), 185–200, at 187, 198.

69. Ibid., 190.

70. See, for example, Chapter 2, "Redefining Death in America, 1968." Also see Susan E. Lederer, *Subjected to Science: Human Experimentation in America before the Second World War* (Baltimore:

Johns Hopkins University Press, 1995). For those who focus on bioethics as a response to post–World War II ethical abuses in human experimentation, Lederer's account of concerns over patients' rights before the war raises the question as to why bioethics should have developed so late when both abuses and concern over those abuses had such a long history.

71. Rothman, *Strangers at the Bedside,* 148, 149.

72. Rothman, "View the First," S11-S12.

73. Cited in Loren Baritz, *The Good Life: The Meaning of Success for the American Middle Class* (New York: Harper and Row, 1989), 256.

74. Rothman, "Human Experimentation and the Origins of Bioethics in the United States," 189. See also, David J. Rothman, "Ethics and Human Experimentation: Henry Beecher Revisited," *New England Journal of Medicine* 317 (November 5, 1987): 1195–99.

75. For example, Andre Hellegers helped to found the Kennedy Institute of Ethics in Georgetown, and Willard Gaylin cofounded the Hastings Center in New York. Other early physicians-scientists include Henry K. Beecher, Theodosius Dobzhansky, Renee Dubos, Leon Kass, and Jay Katz.

76. For example, George Annas, Alexander Capron, William J. Curran, and Charles Fried.

77. Sociologist Renee Fox has written on the significance of the small number of sociologists involved with bioethics. See Renee Fox, "Advanced Medical Technology—Social and Ethical Implications," in *Annual Review of Sociology,* ed. Alex Inkeles, James Coleman, and Neil Smelser (Palo Alto: Annual Reviews, 1976), 2:231–68.

78. Joseph Fletcher, *Morals and Medicine* (Boston: Beacon Press, 1960), 146, 37, xx, 11.

79. Ibid., xv, xvii, 10.

80. Paul Ramsey, *The Patient as Person: Explorations in Medical Ethics* (New Haven: Yale University Press, 1970), xi.

81. Ramsey, *Fabricated Man,* 122–23.

82. Ibid., 135, 138.

83. See Jonsen, "The Birth of Bioethics," S1: "A pioneer was defined as one whose name had appeared in the first edition of the *Bibliography of Bioethics* (1975) and who had continued to work in the field. Some sixty persons made the cut, and of these, forty-two came to Seattle." Jonsen also explained why the thirty-year figure was chosen: "The occasion of this conference was the thirtieth anniversary

of the publication of an article in *Life* magazine, 'They Decide Who Lives, Who Dies' (November 9, 1962). That article told the story of a committee in Seattle whose duty it was to select patients for entry into the chronic hemodialysis program recently opened in that city."

84. Ibid., S4.

85. Willard Gaylin, "Fooling with Mother Nature," *Hastings Center Report* 20, no. 1(1990): 56–60.

Chapter Two: "Leader of Leaders"

1. Statement of Award, 1986, Hastings Center Archives, Garrison, N.Y., formerly at Briarcliff Manor, N.Y. (hereafter cited as HCA).

2. Memo from Daniel Callahan to Board of Directors, February 15, 1991, HCA.

3. Daniel Callahan referred to NIH's 5 percent budget for this ethical program in "Bioethics, Our Crowd, and Ideology," *Hastings Center Report* 26, no. 6 (1996): 3–4, at 3. George Annas referred to NIH's 3 percent budget in his presentation at the Birth of Bioethics Conference, Seattle Washington, September 23–24, 1992.

4. The term *biological revolution* was adopted from molecular biologists-geneticists.

5. This type of analysis is represented by the social commentaries of Jacques Ellul, Herbert Marcuse, Theodore Roszak, Eliot Freidson, and Ivan Illich. For an overview of some of these critiques, see Chapter 1.

6. See Minutes of the Annual Meeting of Members, January 10, 1970, HCA. I use the name the "Hastings Center" anachronistically to avoid an awkwardness in the narrative; also, note that in the narrative, references are made variously to "the institute" and "the center."

7. *Hastings Center Report* 5, no. 3 (1975): opposite 1.

8. Daniel Callahan, "Hastings Center, Institute of Society, Ethics and the Life Sciences," in *The Greenwood Encyclopedia of American Institutions* (1982), vol. 5, s.v. "Research Institutes and Learned Societies." An earlier document has a somewhat different rendition of first interests. See Center for the Study of Value and the Sciences of Man, April 1, 1969, "Report of March 29 Meeting," HCA: "On the areas of interest, it was agreed that, at least initially, the Center could focus on four general fields: population-family planning, medicine, biology and ecology. . . . everyone felt the way should be left open

from the outset for a concern with other areas as well, particularly in the social sciences."

9. Gene I. Maeroff, "The Hastings Center—a Cool Look at Hot Issues," *Change*, February 1979, 12–13.

10. From the bylaws, "Materials Sent to Members, 1969," HCA: "The founders shall include the seventeen persons named below, who were actively involved in the affairs of the Institute prior to its formal incorporation: Henry Beecher, Daniel Callahan, Rene Dubos, John Fletcher, Renee Fox, Willard Gaylin, Martin Golding, Harold P. Green, James Gustafson, Leon Kass, John Maguire, William J. Nagle, Ralph Potter, Paul Ramsey, Alfred Sadler, Blair Sadler, Donald Warwick."

11. "Bioethics and Our Lives: What the Center Does," *Hastings Center Report* 26, no. 1 (1996): special insert.

12. Maeroff, "The Hastings Center—a Cool Look at Hot Issues," 12–13.

13. Jane Stein, "The Bioethicists: Facing Matters of Life and Death," *Smithsonian* 9 (January 1979): 107–15.

14. Ibid., 112.

15. Callahan, "Hastings Center, Institute of Society, Ethics and the Life Sciences," 5:258.

16. Ibid., 259–60.

17. See, "Bioethics and Our Lives: Who Supports the Hastings Center," *Hastings Center Report* 26, no. 1 (1996): special insert, for a list of corporate, foundation, and "roundtable" support (i.e., support from individuals who contribute $1,000 or more annually).

18. Callahan, "Hastings Center," 257–58.

19. Daniel Callahan, "Center For the Study of Social Ethics: A Preliminary Sketch," in folder: Institute of Society, Ethics and the Life Sciences, Inc., 1969, p. 1, Rockefeller Brothers Fund, Rockefeller Archives, Sleepy Hollow, N.Y.

20. "Institute of Society, Ethics and the Life Sciences: A Survey of Goals, Plans and Budgetary Needs," January 1970, in folder: ISELS, 1970, p. 1, Rockefeller Brothers Fund, Rockefeller Archives.

21. Renee Fox, "Bioethics, Our Crowd, and Ideology," *Hastings Center Report* 26, no. 6 (1996): 6–7.

22. Oral history of Daniel Callahan, by M. L. Tina Stevens, Hastings Center, Briarcliff Manor, N.Y., July 1, 1991.

23. Quoted in David Rothman, *Strangers at the Bedside: A History of How Law and Bioethics Transformed Medical Decision Making* (New York: Basic Books, 1991), 209.

24. See Oral History of Daniel Callahan, July 1, 1991, transcript, 7–8.

I think there were two or three things that led to [the] emergence of the field. One was the fact that great technological changes were taking place in medicine, particularly in the 60s—the burst of new technologies—really made a big difference. So we were getting a lot of new things coming along, all of which seemed to carry some new moral dilemma—dilemmas that had not been part of traditional medicine before. The first question was how to respond to . . . new moral problems that were being generated by the new medical technological and biological problems.

I think that the second thing is a cultural point, namely, as far as the sixties there was a great deal of interest in the idea of rights, and the script got translated into: should women . . . have rights, and [should] all sorts of other creatures . . . have rights? Should now patients have rights? Now that cultural idea became very much involved in the doctor-patient relationship and in the institution of medicine and it was kind of a period of saying, "Why, we have been dominated by experts before, we need more democratic participation. Lay people should have rights and powers over against experts, be they doctors or anybody else." So you had two things coming together really, three things, now that I come to think of it. . . .

Say, you had all the medical developments and they were really creating a lot of problems with traditional medical ethics which simply had not been envisioned. And then you have all the cultural and legal developments such as the rise of rights, the anti-establishment criticism of experts—a whole variety of things that came in as a second stream. I think that the third stream probably was that it was a period of creating lots of new organizations. . . .

So I think there were three things that were coming together. People were starting lots of new ventures in those days. And we were one of them. That's essentially why the field came about. And then I happened to be in the right place at the right time, getting interested in these matters.

25. Paul Ramsey, *The Patient as Person: Explorations in Medical Ethics* (New Haven: Yale University Press, 1970), xvi.

26. Robert Veatch, "The Generalization of Expertise," *Hastings Center Studies* 1, no. 2 (1973): 29.

27. "Report on September 26—27 Meeting, 1969," 3, HCA.

28. Memorandum of August 8, 1969, HCA.

29. "Early Fall Notes, Institute for the Study of Ethics and Society," October 1971, HCA. Press and journal coverage favorable to the center continued throughout the decade. See, for example, Kenneth L. Woodward, "The Ethics of Miracles," *Newsweek*, September 19, 1977, 115—18; Dava Sobel, "The Hastings Center: Think before You Act," *RF Illustrated* 4, no. 2 (September 1978): 4—5; Jane Stein, "The Bioethicists: Facing Matters of Life and Death," *Smithsonian* 9 (January 1979): 107—15; Gene I. Maeroff, "The Hastings Center—a Cool Look at Hot Issues," *Change* 11 (February 1979): 12.

30. "Report to the Fellows, January 11—12, 1974," 1—2, HCA.

31. Ibid.

32. "Fellows Meeting, January 12, 1974," 4—5, HCA.

33. See, "Appendix: The Institute as an Activist Organization," in "Report on Institute Activities: 1972—73," HCA. Sometimes use of the word indicated a lack of agreement on which end of the pro- and antitechnology debate the term *activist* was meant to signify. A letter from ethicist John Fletcher, for example, referred to Joseph Fletcher (well known for his strongly held pro-technology views) as an "activist" and "change-oriented" person. Another use of "activist," however, suggested engagement with methods aimed at slowing applications of technological "advance." See John Fletcher to Daniel Callahan, June 13, 1979, HCA.

34. "Appendix: The Institute as an Activist Organization," in "Report on Institute Activities: 1972—1973," HCA. See also, "Early Fall Notes, October, 1971," 3.

35. "Appendix: The Institute as an Activist Organization," 1. See also Minutes of the Fellows Meeting for January 13, 1973, 6.

36. Records indicate that institute representatives participated in the psychosurgery inquiry. See Minutes of the Fellows Meeting for January 13, 1973, HCA. At this meeting, fellows discussed how activist the center should be. Legal counsel reported that the Internal Revenue Code restricted the attempts that could be made to influence legislation. The chair (Gaylin) emphasized that activism did not simply mean legislative activities and made reference to the institute's involvement with the Mississippi psychosurgery case. See 6—8: "A lengthy discussion took place on the desirability of participating in matters such as this. Alternatively, it was suggested, the Institute might confine itself to issuing generalized guidelines and standards worked out by its various research groups. . . . no decisions were reached."

37. "Appendix: The Institute as Activist Organization," 4.

38. "Fellows Meeting, January 12, 1974," 5.

39. "Report 1972–73," 4.

40. Ibid., 3.

41. Joel Margolis, Office of Representative William Steiger, Congress of the United States, House of Representatives, to Daniel Callahan, June 29, 1970, HCA.

42. "May 16, 1975 agenda for the Board of Directors Meeting of June 13, 1975," 2, HCA.

43. See references to this letter quoted in a letter from Harold P. Green to Daniel Callahan, July 20, 1971. See also Sheldon Wolff to Daniel Callahan, July 8, 1971: "I do not wish to leave you with the impression that I felt that the group was negative, but rather a few individuals, perhaps more vocal than myself."

44. Harold Green to Daniel Callahan, July 20, 1971, HCA.

45. Minutes of the Behavior Control Task Force, October 17–18, 1970, HCA.

46. Memo to Daniel Singer, Membership Committee, from Gaylin and Callahan, 1970–71 (otherwise undated), HCA.

47. John Fletcher to Daniel Callahan, July 2, 1970,HCA.

48. Minutes of the Board of Directors Meeting, June 24, 1972, HCA.

49. Paul Ramsey to Daniel Callahan, August 9, 1972, 2, HCA.

50. See "Minutes of the Board of Directors," June 24, 1972, 2, HCA.

51. "Memorandum to the Board of Directors," May 4, 1972, 2, HCA.

52. "Memo to the Board, October 6, 1971," HCA.

53. Ibid.

54. Memo to Daniel Callahan from Willard Gaylin, June 5, 1973, HCA.

55. "Projected Timetable, 1969," HCA.

56. "Memo to the Board of Directors," October 14, 1994, HCA.

57. Oral History of Daniel Callahan, July 1, 1991, transcript, 6.

58. Daniel Callahan's address to the Birth of Bioethics Conference.

59. "Projected Timetable, 1969," HCA.

60. Constance Foshay, HEW, to Daniel Callahan, June 30, 1971, HCA. Ms. Foshay was interpreting the critical comments made by "Dr. Cooper."

61. Minutes of the Fellows Meeting, January 12, 1974, 7, HCA.

62. Daniel Callahan's address to the Birth of Bioethics Conference.

63. Memo from Donald Warwick to Daniel Callahan, November 23, 1973, HCA.

64. Memo from Donald Warwick to Daniel Callahan, December 5, 1973, HCA.

65. Memo from Robert Veatch to Daniel Callahan, November 30, 1973, HCA.

66. Memo from Donald Warwick to Daniel Callahan, April 17, 1974, HCA.

67. Memo from Donald Warwick to Daniel Callahan, December 5, 1973, HCA.

68. Daniel Callahan, "The Emergence of Bioethics," in *Science, Ethics, and Medicine,* ed. H. Tristram Engelhardt Jr. and Daniel Callahan, xv, Foundations of Ethics and Its Relationship to Science, vol. 1 (Hastings-on-Hudson, N.Y., c. 1976).

69. May 16, 1975, Agenda for the June 13, 1975, Board of Directors Meeting, 2, HCA.

70. Ibid.

71. Draft letter from Daniel Callahan to Dr. Fred Bergmann, National Institutes of Health, May 1975, 2, HCA.

72. Ibid., 3, HCA.

73. Ibid., 6, HCA.

74. The June 13, 1975, Board of Directors Minutes, HCA, state that Callahan reported "on what had been regarded . . . as an apparently increasing reluctance of Federal agencies to support projects relating to ethical issues in biomedical science and technology. He indicated that there was reason to believe that the apparent trend may have been arrested or reversed." There is nothing else in the minutes about the backlash.

75. Maeroff, "The Hastings Center—a Cool Look at Hot Issues," 13.

76. Daniel Callahan, *The Tyranny of Survival (and Other Pathologies of Civilized Life)* (New York: Macmillan, 1973), 247, 260.

77. Daniel Callahan, "The Future of the Hastings Center: Reflections and Proposals," November 27, 1990, draft, 2, 3, HCA

78. Ibid. See also p. 4: "We would focus as well on questions of the ends of medicine, changing conceptions of illness and health, and the place of a pursuit of health in individual and societal life. A special goal will be to constantly relate the bioethics projects to the larger issues."

79. Daniel Callahan, "Calling Scientific Ideology to Account," *Society* 33, no. 4 (May–June 1996): 15, 17.

80. Ibid., 19.

81. Daniel Callahan, "Bioethics, Our Crowd, and Ideology," *Hastings Center Report* 26, no. 6 (1996): 3–4, at 3.

Chapter Three: Redefining Death in America, 1968

Portions of this chapter are reprinted from M. L. Tina Stevens, "Redefining Death in America, 1968," *Caduceus: A Humanities Journal for Medicine and the Health Sciences* 11, no. 3 (winter 1995), 207–9, and "What Quinlan Can Tell Kevorkian about the Right to Die," *Humanist* 57, no. 2 (1997): 10–14.

1. "A Way of Dying," *Atlantic Monthly* 199 (January 1957): 53.

2. Editorial, "Life-in-Death," *New England Journal of Medicine* 256, no. 16 (April 18, 1957): 760.

3. John J. Farrell, M.D., "The Right of a Patient to Die," *Journal of the South Carolina Medical Association* 54 (July 1958): 231–32.

4. See, for example, Thomas J. O'Donnell, S.J., "Artificial Resuscitation: A Moral Evaluation," *Georgetown Medical Journal* February 1961, 242–44; Frank J. Ayd Jr., M.D., "The Hopeless Case: Medical and Moral Considerations," *JAMA* 181, no. 13 (September 29, 1962): 1099–1102 (also pp. 83–86); William P. Williamson, M.D., "Life or Death—Whose Decision?" *JAMA* 197, no. 10 (September 5, 1966): 139–41 (also pp. 793–95); Very Rev. Brian Whitlow and Fred Rosner, M.D., "Extreme Measures to Prolong Life," *JAMA* 202, no. 4 (October 23, 1967): 226–28 (also pp. 374–76); J. Russell Elkinton, M.D., "When Do We Let the Patient Die?" *Annals of Internal Medicine* 68, no. 3 (March 1968): 695–700.

5. Pope Pius XII, "Allocution Delivered to the International Congress of Anesthesiologists," November 24, 1957, *Acta Apostolicae Sedis* 49 (November 24, 1957): 1027–33, and "The Prolongation of Life," in *Ethics in Medicine: Historical Perspectives and Contemporary Concerns*, ed. Stanley Joel Reiser, Arthur J. Dyck, and William J. Curran (Cambridge, Mass.: MIT Press, 1977), 501–4.

6. As cited in Ayd, "The Hopeless Case: Medical and Moral Considerations," 1102. See also O'Donnell, "Artificial Resuscitation: A Moral Evaluation," 242–44. The archbishop of Canterbury (Cosmo Lord Lang) also concluded that "cases arise in which some means of shortening life may be justified," cited in Joseph Fletcher, "The Patient's Right to Die," *Harper's Magazine*, October 1960, 142.

7. Fletcher, "The Patient's Right to Die," 140, 141, 143.

8. Cited in Ayd, "The Hopeless Case: Medical and Moral Considerations," 1099–1100.

9. Williamson, "Life or Death—Whose Decision?" 139.

10. Ibid., 141.

11. Ibid.

12. See, for example, "Needless Transplants?" *Newsweek*, March 2, 1964, 74, quoting Dr. J. Russell Elkinton stating that while the transplant record "augurs well for the clinical future, from the point of view of the clinical present [it] is less than rosy—it is black." See also "Year of the Transplant," *Newsweek*, February 10, 1964, 52, quoting a researcher as saying, "I have no words to describe my opposition—scientifically and even morally—to these transplants."

13. John H. Kennedy, "Cardiac Transplantation—a Current Appraisal," *JAMA* 203, no. 10 (March 4, 1968): 172–73.

14. Editorial, "What and When is Death?" *JAMA* 204, no. 6 (May 6, 1968): 219–20.

15. For a historical examination of definitions of death preceding 1968, see Martin S. Pernick, "Back From the Grave: Recurring Controversies over Defining and Diagnosing Death in History," in *Death: Beyond Whole-Brain Criteria*, ed. Richard M. Zaner (Boston: Kluwer Academic Publishers, 1988), 17–74.

16. "A Definition of Irreversible Coma. Report of the Ad Hoc Committee of the Harvard Medical School to Examine the Definition of Brain Death," *JAMA* 205, no. 6 (August 5, 1968): 337–40; see also, Henry K. Beecher, M.D., "Ethical Problems Created by the Hopelessly Unconscious Patient," *New England Journal of Medicine* 278, no. 26 (June 27, 1968): 1425–30.

17. For an account of the ad hoc committee's redefinition of death which differs from the analysis offered here, see David Rothman, *Strangers at the Bedside: A History of How Law and Bioethics Transformed Medical Decision Making* (New York: Basic Books, 1991). For an account of the redefinition of death compatible with the one offered here, see Mita Giacomini, "From Death Defying to Death Defining: Technological Imperatives and the Definition of Brain Death in 1968" (M.A. thesis, History of Health Sciences, University of California at San Francisco, 1992).

18. This assumes, in part, that brain dead people "suffered" while on the respirator—something that the committee never teases out.

19. Lyman A. Brewer, M.D., "Cardiac Transplantation: An Ap-

praisal," *JAMA* 205, no. 10 (September 2, 1968): 101. See also Kennedy, "Cardiac Transplantation—a Current Appraisal," 172.

20. "Ethical Problems in Organ Transplantation," *Annals of Internal Medicine* 67 (September 1967): 33.

21. See Brewer, "Cardiac Transplantation," 101–2.

22. See "Surgery and Show Biz," *Newsweek*, January 15, 1968, 49, and "Surgical Show Biz," *Nation*, January 22, 1968, 100. See Brewer, "Cardiac Transplantation," regarding how some members of the medical profession disapproved of the publicity.

23. "A Plea for a Transplant Moratorium," *Science News* 93 (March 16, 1968): 256.

24. For an analysis that underscores the role of the EEG in redefining death, see Giacomini, "From Death Defying to Death Defining."

25. "Definition of Irreversible Coma," 87.

26. Gordon L. Snider, "Historical Perspective on Mechanical Ventilation: From Simple Life Support System to Ethical Dilemma," *American Review of Respiratory Diseases* 140 (August 1989): S2-S7, and *Collier's Encyclopedia* (New York: Macmillan Educational, 1990), 13:296; S. J. Somerson, "Historical Perspectives on the Development and Use of Mechanical Ventilation," *Aana Journal* 60, no. 1 (February 1992): 83–84. See also Fletcher, "The Patient's Right to Die," which discusses terminating artificial respiration.

27. Thomas E. Starzl, *The Puzzle People: Memoirs of a Transplant Surgeon* (Pittsburgh: University of Pittsburgh Press, 1992), 148.

28. Ibid., 148–49.

29. James Z. Appel, M.D., "Ethical and Legal Questions Posed by Recent Advances in Medicine," *JAMA* 205, no. 7 (August 12, 1968): 102.

30. D. D. Rutstein, "The Ethical Design of Human Experiments," *Daedalus* 98 (spring 1969): 526.

31. "An Appraisal of the Criteria of Cerebral Death—a Summary Statement—a Collaborative Study," *JAMA* 237, no. 10 (March 7, 1977): 982–86, at 982.

32. "Death, When Is Thy Sting?" *Newsweek*, August 19, 1968, 54.

33. "Redefining Death," *Newsweek*, May 20, 1968, 68.

34. "Pioneering Heart Transplant Surgeon Retires," *Stanford Daily*, January 25, 1993, 1.

35. "A Definition of Irreversible Coma," 87.

36. Ibid.

37. "If this position is adopted by the medical community, it can form the basis for change in the current legal concept of death. No statutory change in the law should be necessary since the law treats this question essentially as one of fact to be determined by physicians. The only circumstance in which it would be necessary that legislation be offered in the various states to define 'death' by law would be in the event that great controversy were engendered surrounding the subject and physicians unable to agree on the new medical criteria." Ibid., 339.

38. "Definitions of Death," *Scientific American*, December, 1971, 40.

39. "Symposium Hears Transplant Plea," *New York Times*, September 9, 1968, 23.

40. See for example, George P. Fletcher, J.D., "Legal Aspects of the Decision Not to Prolong Life," *JAMA* 203, no. 1 (January 1, 1968): 122.

41. "When Are You Really Dead?" *Newsweek*, December 18, 1967, 87.

42. "What Is Life? When Is Death?" *Time*, May 27, 1966, 78.

43. John D. Arnold, M.D., Thomas F. Zimmerman, Ph.D., and Daniel C. Martin, M.D., "Public Attitudes and the Diagnosis of Death," *JAMA* 206, no. 9 (November 25, 1968): 1949–54. See also Editorial, "When Do We Let the Patient Die?" *Annals of Internal Medicine* 68, no. 3 (March 1968): 698.

44. Arnold, Zimmerman, and Martin, "Public Attitudes and the Diagnosis of Death," 1954.

45. Ulys H. Yates, "Transplantation: Today and Tomorrow," *Today's Health*, April 1968, 37.

46. "State Seeking to Define Transplant Donor Death," *New York Times*, November 28, 1968, 48.

47. "Definition of Death," *Science Digest*, March 1969, 77. For examples of popular discussion of transplantation and the redefinition of death, see also "Redefining Death," *Newsweek*, May 20, 1968, 68; Lawrence Lader, "Who Has the Right to Live?" *Good Housekeeping*, June 1968, 85, 144–48; "Death, When Is Thy Sting?"; Leonard A. Stevens, "When Is Death?" *Reader's Digest*, May 1969, 225–32. For other examples of organized medicine's interpretation of same for popular journalistic consumption, see Russell S. Fisher, M.D., "Let the Dead Help the Living," *Today's Health*, April 1969, 88–87, and Allon Ochsner, M.D., "Medical Breakthroughs You Can Expect in 10 . . . 25 . . . 50 Years," *Today's Health*, April 1973, 44–49.

48. J. Russell Elkinton, M.D., "When Do We Let the Patient Die?" *Annals of Internal Medicine* 68, no. 3 (March 1968): 698.

49. "The Right of a Patient to Die Stressed to Medical Graduates," *New York Times*, June 2, 1968, 57.

50. "Symposium Hears Transplant Plea," *New York Times*, September 9, 1968, 23.

51. "Donors for Organ Transplants—Letter to the Editor," *JAMA* 207, no. 13 (March 31, 1969): 2439.

52. "Physicians Adopt a Code on Death," *New York Times*, August 10, 1968, 25. See also Rutstein, "The Ethical Design of Human Experimentation," 523–41, and J. F. Toole, "The Neurologist and the Concept of Brain Death," *Perspectives in Biology and Medicine* 14 (1971): 599–606, for other criticism of brain death criteria from physicians.

53. Editorial, "Redefining Death," *New York Times*, August 8, 1969, 32.

54. Ad Hoc Committee of the American Electroencephalographic Society, "Cerebral Death and the Electroencephalogram," *JAMA* 209, no. 10 (September 8, 1969): 1505–9.

55. S. Drucker, Letter to the Editor, *New York Times*, September 1, 1968, E11.

56. Fred Anderson, "Death and the Doctors: Who Will Decide Who Is to Live?" *New Republic* 160 (April 19, 1969): 9–10.

57. See, for example, *Tucker v. Lower*, No. 831 (Richmond, Va., L. & Eq. Ct., May 23, 1972); *Commonwealth v. Golston*, 373 Mass. 249, 366 N. E. 2d 744 (1977); *Lovato v. District Court*, 198 Colo. 419, 601 P. 2d 1072 (1979); *People v. Saldana*, 47 Cal. App. 3d 954, 121 Cal. Rptr. 243 (1975); and, *State v. Fierro*, 124 Ariz. 182, 603 P. 2d 74 (1974).

58. Paul Speegle, "BAR Talk," *Recorder* 96, no. 92 (May 13, 1974): 1.

59. "Dead Girl Breathes in Nick of Time," *San Francisco Examiner*, August 5, 1974, 9.

60. See, for example, "A Plea for a Transplant Moratorium," 256.

61. "In Brief," *Hastings Center Report* 2, no. 6 (1972): 16.

62. Quoted in "When Are You Really Dead?" 87.

63. Quoted in Lader, "Who Has the Right to Live?" 146.

64. Hans Jonas, "Philosophical Reflections on Experimenting with Human Subjects," *Daedalus* 98, no. 2 (spring 1969): 219–47, at 243–44.

65. Paul Ramsey, *The Patient as Person: Explorations in Medical Ethics* (New Haven: Yale University Press, 1970), 112.

66. Henry K. Beecher to Daniel Callahan, April 30, 1970, HCA.

67. Daniel Callahan to Henry K. Beecher, April 27, 1970, HCA.

68. Henry K. Beecher to Daniel Callahan, April 30, 1970, HCA.

69. Henry K. Beecher, "Scarce Resources and Medical Advancement," *Daedalus,* Spring 1969, 305.

70. Henry K. Beecher to Daniel Callahan, February 25, 1971; Daniel Callahan to Henry K. Beecher, March 2, 1971; Henry K. Beecher to Daniel Callahan, March 15, 1971, HCA.

71. Henry K. Beecher to Daniel Callahan, April 28, 1971, HCA. Beecher's summary is as follows:

1) I braved the wrath of the medical and scientific world at the Brook-Lodge conference in March, 1965. Fifteen months later I published that material in the *New England Journal of Medicine* article, "Ethics and Clinical Research." This evoked an even greater outpouring of scorn. During those difficult days I remained quite unrepentant. Well, that is pretty much water over the dam (almost everybody speaks to me once again and some of the die-hards have even admitted they were wrong in their attitude toward what I was trying to do. For example, F. J. Ingelfinger of the *New England Journal of Medicine* and, as you know, the *Lancet,* editorially, etc.).

2) My next endeavor was to ask the Dean at Harvard if I could get together a group to study irreversible coma. He agreed. From a period of utter chaos at our first meeting until the last meeting (with the five drafts I wrote of our final report in hand) we had unanimity. This, I think, has served a useful purpose.

3) My next effort was to state and defend my position that "irreversible coma" is brain death and that the easily diagnosed brain death is death indeed. That view has been quite generally accepted by physicians, as far as I can find.

4) I have been trying earnestly to get concurrence of opposing groups, lawyers, theologians, philosophers, so far without much success. I have not given up hope.

5) Just now I am working on an article with the title, "Fear of Death as a Cause of Death." (An abundance of material is available on this.)

72. Daniel Callahan to Henry K. Beecher, May 13, 1971; Henry K. Beecher to Daniel Callahan, May 19, 1971, HCA.

73. See also Henry K. Beecher to Daniel Callahan, February 25, 1971; Daniel Callahan to Henry K. Beecher, March 2, 1971; and

Henry K. Beecher to Daniel Callahan, April 28, 1971, HCA, for further indications of tensions between Beecher and other members of the death group.

74. "Refinements in Criteria for the Determination of Death: An Appraisal," Report by the Task Force on Death and Dying of the Institute of Society, Ethics and the Life Sciences, *JAMA* 221, no. 1 (July 3, 1972): 52, 51, 52.

75. "Refinements," 48, 52.

76. On the matter of "harvesting" organs, see Willard Gaylin, "Harvesting the Dead," *Harper's*, September 1974, 22–30, and William May, "Attitudes toward the Newly Dead," *Hastings Center Studies* 1, no. 1 (1973): 3–13.

77. For examples of the effort to come to terms with continuing difficulties over defining death *after* the 1968 Harvard Committee criteria, see Henry K. Beecher, M.D., Raymond Adams, M.D., and William H. Sweet, M.D., "Procedures for the Appropriate Management of Patients Who May Have Supportive Measures Withdrawn," *JAMA* 209, no. 3 (July 21, 1969): 405; Henry K. Beecher, M.D., "After the 'Definition of Irreversible Coma,'" *New England Journal of Medicine* 281, no. 19 (November 6, 1969): 1070–72; "An Appraisal of the Criteria of Cerebral Death—a Summary Statement—a Collaborative Study," 982–86; Peter Mcl. Black, M.D., "Brain Death," *New England Journal of Medicine* 299, no. 8 (August 24, 1978): 393–401; Paul A. Byrne, M.D., Sean O'Reilly, M.D., FRCP, Paul M. Quay, S.J., Ph.D., "Brain Death—an Opposing View," *JAMA* 242, no. 18 (November 2, 1979): 1985–90; Robert Veatch, "Defining Death: The Role of Brain Function," *JAMA* 242, no. 18 (November 2, 1979): 2001–2; Millard Bass, Letter to Editor, *JAMA* 242, no. 17 (October 26, 1979): 1850. See also "A Statutory Definition of the Standards for Determining Human Death: An Appraisal and a Proposal," *University of Pennsylvania Law Review* 121 (November 1972): 87–118.

78. See Rothman, *Strangers at the Bedside*, 156–65.

Chapter Four: "Sleeping Beauty"

Portions of this chapter are reprinted from M. L. Tina Stevens, "The Quinlan Case Revisited: A History of the Cultural Politics of Medicine and the Law," *Journal of Health Politics, Policy and Law* 21, no. 2 (summer 1996): 347–66, and "What *Quinlan* Can Tell Kevorkian about the Right to Die," *Humanist* 57, no. 2 (1997): 10–14.

1. Adapted from *Karen Ann: The Quinlans Tell Their Story*, by

Joseph Quinlan and Julia Quinlan with Phyllis Battelle (Garden City, N.Y.: Doubleday and Company, 1977).

2. For views of the Quinlan case that differ from the interpretation offered here, see Peter G. Filene, *In The Arms of Others: A Cultural History of the Right-to-Die in America* (Chicago: Ivan R. Dee, 1998), and David J. Rothman, *Strangers at the Bedside: A History of How Law and Bioethics Transformed Medical Decision Making* (New York: Basic Books, 1991).

3. See, for example, Ellen Goodman, "Living and Dying: Are They Medical Concerns?" Op-Ed, *Washington Post*, June 15, 1976, A-21; Editorial, "Life, Death, Comas and Courts," *Chicago Tribune*, June 16, 1985, sec. 5, 2.; and, "Karen Quinlan Dies; 10 Year Coma Aided Right-To-Die Rulings," *New Orleans Times Picayune*, June 12, 1985, 1.

4. Quinlan, Quinlan, and Battelle, *Karen Ann: The Quinlans Tell Their Story*, 161.

5. Ibid., 223.

6. "Who Was Karen Quinlan?" *Newsweek*, November 3, 1975, 60.

7. Ibid.

8. "Probe Links Drugs to Coma," *Chicago Tribune*, September 20, 1975, sec. 1 (N), 7; "Drugs and Alcohol Linked to Coma Case," *Washington Post*, September 20, 1975, A6.

9. "Karen Believed Beaten before Going into Coma," *Los Angeles Times*, December 17, 1975, sec. 1, 6; "Probe into Coma of Karen Quinlan Ends," *San Francisco Chronicle*, February 25, 1976, 7.

10. "Foul Play Probed in Quinlan Case," *Washington Post*, December 12, 1975, A8; "Probe Rules Out Foul Play as Cause of Quinlan Coma," *Washington Post*, February 25, 1976, A6.

11. Quinlan, Quinlan, and Battelle, *Karen Ann: The Quinlans Tell Their Story*, 222.

12. Donald Kirk, "Parents' Plea Is Denied," *Chicago Tribune*, November 11, 1975, 1.

13. Charles M. Whelan, "Karen Ann Quinlan: Patient or Prisoner?" *America*, November 22, 1975, 346.

14. "Right-to-Die Decision Could Have Broad Effects," *Chicago Tribune*, October 26, 1975, sec. 1, 14.

15. See B. D. Colen, *Karen Ann Quinlan: Dying in the Age of Eternal Life* (New York: Nash Publishing, 1976).

16. Ibid., 25.

17. "Right-to-Die Decision Could Have Broad Effects," sec. 1, 14.

18. Charles McCabe, "Karen Ann," *San Francisco Chronicle*, October 29, 1975, 41.

19. "Quinlan Case Goes to Trial," *New Orleans Times Picayune*, October 20, 1975, sec. 1, 16.

20. *San Francisco Chronicle*, September 16, 1975, 1; *Washington Post*, September 16, 1975, 1.

21. Quinlan, Quinlan, and Battelle, *Karen Ann: The Quinlans Tell Their Story*, 177; "Physician Testifies in Mercy Death Plea," *Los Angeles Times*, October 21, 1975, sec. 1, 5.

22. "Doctor Fights Call for Karen's Death," *Chicago Tribune*, October 21, 1975, sec. 1, 3.

23. "Fate of Miss Quinlan Will Be Decided Soon," *New Orleans Times Picayune*, October 28, 1975, sec. 1, 2.

24. "Lawyer Argues against Mercy Killing," *Los Angeles Times*, October 28, 1975, sec. 1, 8.

25. "A Right to Die?" *Newsweek*, November 3, 1975, 59.

26. Letter to the Editor, "Karen Quinlan," *Chicago Tribune*, November 1, 1975, sec. 1, 8.

27. "A Life in the Balance," *Time*, November 3, 1975, 61.

28. "Karen Ann Quinlan: *Journal* Readers Speak Out," *Ladies Home Journal*, January 1977, 112.

29. "Quinlan Family Gets Advice and Sympathy in Mail," *New York Times*, October 26, 1975, 46.

30. Ibid.

31. Joan Kron, "The Girl in the Coma," *New York*, October 6, 1975, 31.

32. "A Right to Die?" 67.

33. "Current Comment: Turning Off the Machine," *America*, October 11, 1975, 197.

34. See, for example, "The Quinlan Tragedy," *St. Louis Post-Dispatch*, November 13, 1975, 3B; "Karen Anne's 'Life,'" *San Francisco Examiner and Chronicle*, October 19, 1975, sec. B, 2.

35. Georgie Anne Geyer, "A Question of Life and Death," *Los Angeles Times*, October 27, 1975, sec. 2, 7.

36. Letter to the Editor, *Los Angeles Times*, October 30, 1975, sec. 2, 6.

37. Peter Steinfels, "The Quinlan Decision," *Commonweal*, December 5, 1975, 584.

38. "Quinlan Family Gets Advice and Sympathy in the Mail," 46.

39. Quinlan, Quinlan, and Battelle, *Karen Ann: The Quinlans Tell Their Story*, 234. One sympathizer, inspired by Karen's tragedy, was moved to write poetic verse she entitled "Karen Ann." The first four lines read: "They hold me here / between two worlds. / The

flesh leaves my hollow bones / Even my breath does not belong to me." Caryl Porter, "Karen Ann," *Christian Century*, January 21, 1976, 45.

40. "Newsmakers," *Newsweek*, May 3, 1976, 46.

41. Letter to the Editor, "On Karen Quinlan," *Milwaukee Journal*, June 23, 1976, sec. 1, 14.

42. "Quinlan Case," *Washington Post*, November 2, 1976, 2.

43. George Will, "The Slippery Slope of 'Meaningful Life,'" *Washington Post*, November 14, 1975, A19.

44. Lonnie R. Bristow, "A Sigh of Relief over Quinlan Ruling," *Los Angeles Times*, December 4, 1975, sec. 2, 7.

45. "The Quinlan Quagmire," *New York Times*, November 11, 1975, 30. See also, for example, "Decision for Life—and a Living Death," *San Francisco Chronicle*, November 16, 1975, Sunday Punch sec., 1; Thomas A. Shannon, "A Triumph of Technology," *Commonweal*, December 5, 1975, 589.

46. "Pulling the Plug" . . . Is It Murder or Mercy?: An In-Depth Survey of American Women," *Ladies Home Journal*, March, 1976, 98–99.

47. Rothman, *Strangers At the Bedside*, 238, noted that, "In 1977, 66 percent of respondents to a poll by Lou Harris approved the proposition [that families "ought to be able to tell doctors to remove all life-support services and let the patient die"] and 15 percent were undecided; four years later, 73 percent approved and only 4 percent were undecided."

48. See, for example, Marquis Childs, "The Quinlan Tragedy," *St. Louis Post-Dispatch*, November 13, 1975, 3B; "On Being Allowed to Die," *Humanist* 36, no. 1 (1976): 17; Letters to the Editor, "Karen Quinlan: Legal, Medical and Human Needs," *Washington Post*, November 18, 1975, A26; M. D. Snyder, Letter to the Editor, *Los Angeles Times*, October 30, 1975, sec. 2, 6; Virginia M. Huddleston, Letter to the Editor, *Los Angeles Times*, October 30, 1975, sec. 2, 6; Roy T. Kobayashi, Letter to the Editor, *Los Angeles Times*, October 30, 1975, sec. 2, 6.

49. "Court Rules Dad May Let Karen Die," *Chicago Tribune*, April 1, 1976, 1; Letter to the Editor, *Christian Century*, April 7, 1976, 342.

50. Dick Harbottle, Letter to Editor, *Los Angeles Times*, October 30, 1975, sec. 2, 6.

51. Letter to Editor, "The Quinlan Case," *Washington Post*, September 23, 1975, A21.

52. "Parents' Death Plea Is Denied," *Chicago Tribune*, November 11, 1975, 1.

53. See, for example, Lawrence H. Eldredge, "Does Stricken Girl Have the Right to Die? Law Says Yes," Op-Ed, *Los Angeles Times*, October 28, 1975, sec. 2, 5.

54. "Quinlans' Death Request Denied," *New Orleans Times Picayune*, November 11, 1975, 1.

55. See, for example, Letter to the Editor, "Karen Quinlan: Legal, Medical and Human Needs," A26; George F. Will, "Affirmation on Life," *San Francisco Chronicle*, November 14, 1975, 44.

56. Whelan, "Karen Ann Quinlan: Patient or Prisoner?" 346–47; Letter to Editor, *Los Angeles Times*, November 18, 1975, sec. 2, 6.

57. Letter to the Editor, *Los Angeles Times*, November 18, 1975, sec. 2, 6.

58. Editorial, "There is a Time to Die," *Chicago Tribune*, April 3, 1976, sec. 1, 10.

59. Goodman, "Living and Dying: Are They Medical Concerns?" A21.

60. Colen, *Karen Ann Quinlan: Dying in the Age of Eternal Life*, 50.

61. *Los Angeles Times*, November 11, 1975, 1; *Washington Post*, November 11, 1975, 1.

62. Quinlan, Quinlan, and Battelle, *Karen Ann: The Quinlans Tell Their Story*, 159.

63. "Fate of Miss Quinlan Will Be Decided Soon," sec. 1, 2; "Parents Ask Court to End Woman's Life," *San Francisco Chronicle*, September 16, 1975, 1. See also Quinlan, Quinlan, and Battelle, *Karen Ann: The Quinlans Tell Their Story*, 217.

64. B. D. Colen, "Court Rule Asked on Life, Death," *Los Angeles Times*, October 19, 1975, sec. 4, 4.

Distressed by the lower court's decision, Timothy Tondreault of California wrote that "the machines have finally won their first landmark decision." Letter to Editor, *Los Angeles Times*, November 18, 1975, sec. 2, 6. According to a *New York Times* editorial, in refusing to rule in favor of Mr. Quinlan, Judge Muir had "refused the opportunity to help bring common law into step with the amazing new capabilities of medical science." Editorial, "The Quinlan Quagmire," 30. The editor of the *San Francisco Chronicle* was dispirited by the court's failure to relieve humankind from this dilemma. "Nothing has been done," he lamented, " . . . to clarify the question, ever more insistently arising in these days of advancing medical technology, of when to pull the plug on the mechanical respirator." Editorial, "Decision for Life—and a Living Death," *San Francisco Chronicle*, November 16, 1975, Sunday Punch section, 1.

65. *U.S. News and World Report*, November 24, 1975, 31.

66. "After almost a year of lying helpless in an irreversibly comatose and vegetative state," exulted the editors of *America*, " . . . Karen may at last be delivered from the respirator that has become her prison"; "Karen Ann Quinlan and the Right to Die," *America*, April 17, 1976, 327. The *Chicago Tribune* was hopeful that the decision would "provide some legal precedent to protect similar patients . . . from becoming victims of the very medical and technological advances that were intended to help them"; Editorial, "There Is a Time to Die," *Chicago Tribune*, April 3, 1976, sec. 1, 10. For one *New York Times* editorial board member, Karen's parents had led a revolt against technology; Harry Schwartz, "On Medical Progress," *New York Times*, April 13, 1976, 33. The *Milwaukee Journal* praised the ruling for being humane and hailed the court as "courageous" in stating that "law, equity and justice must not . . . quail and be helpless in the face of modern technological marvels presenting questions hitherto unthought of"; Editorial, "A Humane Decision," *Milwaukee Journal*, April 2, 1976, sec. 1, 12.

67. Editorial, "Karen Quinlan," *National Review*, April 30, 1976, 438.

68. William F. Hyland and David S. Baime, "In Re Quinlan: A Synthesis of Law and Medical Technology," *Rutgers Camden Law Journal* 8 (1976): 37.

69. Donald Collester Jr., "Death, Dying and the Law: A Prosecutorial View of the Quinlan Case," *Rutgers Law Review* 30 (1977): 304.

70. Colen, *Karen Ann Quinlan: Dying in the Age of Eternal Life*, 68–69.

71. From *McCall's* adaptation of B. D. Colen's *Karen Ann Quinlan: Dying in the Age of Eternal Life*, September 1976, 50.

72. News coverage served to heighten anxiety by reporting on other individuals in situations similar to Karen's. See, for example, "'Plug Pulled' on Maryjane," *New Orleans Times Picayune*, November 18, 1975, sec. 1, 2; "I Did Right to 'Pull Plug' on Husband Widow Says," *Chicago Tribune*, October 28, 1975, sec. 1, 5. One such story was the case of sixteen-year-old Maryjane Dahl, who was attached to a respirator after she lapsed into a coma in late October 1975. On November 2, alerted by a warning buzzer, nurses rushed

into Maryjane's room to find that her respirator had been unplugged. One nurse believed that she had seen members of the patient's family leaving the hospital room. "A Young Girl's Curious Death," *Newsweek*, November 17, 1975, 75. Under the heading, "Another Karen Lingers," the *New York Times* told readers the story of Karen Vikingstad. At the age of fifteen this Karen mysteriously fell into a coma one morning. Never specifically mentioning whether Ms. Vikingstad was on or had ever been placed on mechanical ventilation, the account simply noted that her brain was irreversibly damaged due to lack of oxygen and for two years had been kept alive by "space age technology." Mrs. Vikingstad was quoted as saying that if she had known two years ago what she knew now, she "would have pulled every one of those plugs" that kept her daughter alive. "What can I do now?" she implored. "Can I stop pouring the food into the tube in her stomach and let her slowly starve to death?" *New York Times*, September 25, 1975, 91.

Letters to editors told of similar tragedies. In the *Ladies Home Journal* a parent wrote of the son who at age fourteen suffered a brain stem accident. This anonymous letter writer "regretted the doctors keeping his (son's) body alive with machines when God would have allowed him to die." At the age of seventeen, the victim's prognosis was to live another fifty or sixty years of a "non-life." For this parent, the seventeen-year-old's predicament was an example of the "horrible penalties modern medicine is able to impose upon mankind." "Karen Ann Quinlan: *Journal* Readers Speak Out," 112.

73. "Quinlan Family Gets Advice and Sympathy in Mail," 46.

74. Letter to the Editor, *Newsweek*, November 17, 1975, 4.

75. William F. Buckley, "Avoiding the Quinlan Syndrome," *Los Angeles Times*, November 11, 1975, sec. 2, 5.

76. Childs, "The Quinlan Tragedy," 3B.

77. "Quinlan Alive One Year after Leaving Respirator," *Washington Post*, April 1, 1977, sec. Metro, C6.

78. Ibid.

79. "Karen Ann Quinlan: *Journal* Readers Speak Out," 112.

80. Gregory Gelfand, "Living Will Statues: The First Decade," *Wisconsin Law Review*, no. 5 (1987): 737–822.

81. *In the Matter of Karen Quinlan: The Complete Legal Briefs, Court Proceedings and Decision in the Superior Court of New Jersey* (N.p.: University Publications of America, 1975).

82. Ibid., 214–18.

83. *In Re Quinlan* 70 N.J. 10, pp. 49–52.

84. Editorial, "Doctor's Dilemma," *San Francisco Chronicle,* September 21, 1975, Sunday Punch section, 1.

85. Backers of transplantation research promoted the criteria to facilitate obtaining organs from the bodies of artificially ventilated people whose hearts were still beating—and viable for transplantation—even though their brains had ceased to function. See Chapter 3.

86. "The Quinlan Decision," *Newsweek,* November 24, 1975, 107. Even with these stringent guidelines, mistakes were sometimes made and people were pronounced dead who later were determined to be alive. See Chapter 3.

87. See Henry K. Beecher, Raymond Adams, and William H. Sweet, "Procedures for the Appropriate Management of Patients Who May Have Supportive Measures Withdrawn," *JAMA* 209, no. 3 (July 21, 1969): 495.

88. Quoted in Colen, *Karen Quinlan: Dying in the Age of Eternal Life,* 89. See also H. A. H. van Till—D'Aulnis de Bourouill, "Diagnosis of Death in Comatose Patients under Resuscitation Treatment: A Critical Review of the Harvard Report," *American Journal of Law and Medicine* 2, no. 1 (summer 1976): 5: "if other American jurisdictions follow *Quinlan* in its decision to permit the termination of resuscitation treatment of certain comatose patients under appropriate circumstances, physicians will have less need to find logic-stretching ways to declare such patients dead in order to be permitted to cease such treatment."

89. Glen Affleck and Agnes Thomas, "The Edelin Decision Revisited: A Survey of the Reactions of Connecticut's OB/GYNs," *Connecticut Medicine* 41, no. 10 (October 1977): 637.

90. Carol Altekruse Berger and Patrick Berger, "The Edelin Decision," *Commonweal,* April 25, 1975, 76–78.

91. Affleck and Thomas, "The Edelin Decision Revisited: A Survey of the Reactions of Connecticut's OB/GYNs," 637–38.

92. "Edelin Supported," *New England Journal of Medicine* 292, no. 13 (March 27, 1975): 705.

93. "The Edelin Trial Fiasco," *New England Journal of Medicine* 292, no. 13 (March 27, 1975): 697.

94. "A Right to Die?" 59. See also, "Other Karen Quinlan Cases Have Never Reached Court," *New York Times,* November 2, 1975, sec. 4, 9.

95. See for example, "Lawyer Argues against Mercy Killing," sec. 1, 8; "The Right to Live—or Die," *Time,* October 27, 1975, 41; "A

Malpractice Issue Sparked Coma Case," *New York Times,* September 28, 1975, 1, and "A Right to Die?" 58.

96. Editorial, "Karen Anne's 'Life,'" sec. B, 2; "Father Petitions Court to Let Daughter Die," *Los Angeles Times,* September 17, 1975, sec. 1, 19.

97. *In the Matter of Karen Quinlan, An Alleged Incompetent,* 70 N.J. 10, 355 A. 2d 647 at 20. Karen's court-appointed guardian also acknowledged that it was his "impression that Karen Quinlan was dead, in the sense of 'brain dead'"; *In the Matter of Karen Quinlan, The Complete Legal Briefs, Court Proceedings, and Decision in the Superior Court of New Jersey,* 196.

98. See, "Focus in Quinlan Coma Case Has Shifted," *New York Times,* October 19, 1975, 83; and *In the Matter of Karen Quinlan, An Alleged Incompetent* 137 N.J. Super. 227, 348 A. 2d 801.

99. See for example, "N.J. Case Raises Issue of When Life Ends," *New York Times,* September 21, 1975, sec. 4, 7; and "Quinlans to Appeal Court's Ruling," *New Orleans Times Picayune,* November 17, 1975, sec. 1, 1; "Between Life and Death," *Time,* September 29, 1975, 59; "Byrne Would Back a Bill Covering Terminally Ill," *New York Times,* September 25, 1975, 1.

100. *In the Matter of Karen Quinlan, An Alleged Incompetent,* 70 N.J. 10, 355 A. 2d 647 at 20.

101. "Girl in Coma Not Legally Dead, Says Expert," *New York Times,* October 8, 1975, 43.

102. For examples of effort to clarify, see "A Life in the Balance," 57. For example of continuing misunderstanding, see "Karen Quinlan's Coma," *Christian Century,* October 22, 1975, 916. Editorial, "A Humane Decision," *Milwaukee Journal,* April 2, 1976, sec. 1, 12; "Quinlans to Appeal Court's Ruling," sec. 1, 1; "Quinlan Ruling to Be Appealed," *Washington Post,* November 17, 1975, A4; and Marvin Kohl, "On Death, Dying, and the Karen Quinlan Case," *Humanist* 36, no. 1 (1976): 16. Cf. Tabitha Powledge and Peter Steinfels, "Following the News on Karen Quinlan—a Media Watch on a Landmark Case," *Hastings Center Report* 5, no. 6 (1975): 5.

103. "Karen Ann Quinlan; Coma Prompted Landmark Ruling," *Boston Globe,* June 12, 1985, 63.

104. Editorial, "The Living Dead," *New York Times,* October 12, 1975, 12. See also, for example, Editorial, "A Matter of Life and Death," *Washington Post,* September 30, 1975, A16.

105. "'Right to Die' Is Agony for Many," *New Orleans Times Picayune,* November 3, 1975, sec. 1, 10.

106. Geyer, "A Question of Life or Death," sec. 3, 7; Whelan, "Karen Ann Quinlan: Patient or Prisoner?" 346.

107. "The Right to Live—or Die," 40.

108. "Other Karen Quinlan Cases Have Never Reached Court," sec. 4, 9. See also, Melvin D. Levine, M.D., "Disconnection: The Clinician's View," *Hastings Center Report* 6, no. 1 (1976): 11.

109. See, for example, "Right-to-Die Decision Could Have Broad Effects," sec. 1, 14; "Area Doctors Decide Crucial Cases," *Washington Post*, November 12, 1975, A19.

110. Thomas C. Oden, "A Cautious View of Treatment Termination," *Christian Century*, January 21, 1976, 42.

111. Quinlan, Quinlan, and Battelle, *Karen Ann: The Quinlans Tell Their Story*, 161.

112. "3 Testify Karen Quinlan Can Never Function Again," *New York Times*, October 24, 1975, 41.

113. *In the Matter of Karen Quinlan, An Alleged Incompetent*, 70 N.J. 10, 355 A. 2d 647 at 654.

114. Symposium, "Is There a Right to Die?" *Columbia Journal of Law and Social Problems* 489 (1976): 515–19.

115. "3 Testify Karen Quinlan Can Never Function Again," 41.

116. Rothman, *Strangers at the Bedside*, 229.

117. "A Malpractice Issue Sparked Coma Case," 1.

118. *In the Matter of Karen Quinlan 137 N.J. Super. 227, 348 A. 2d 801*, 252 and 259.

119. "'Right to Die' Case: Will Anything Change?" *U.S. News and World Report*, November 24, 1975, 31.

120. In a similar vein, the notoriety that the act of "pulling the plug" had gained because of the Quinlan case was viewed as a possible inhibiting factor: "Fear can become a factor when decisions must be made in public. It gets very sticky when the patient is on a lot of life-support apparatus—it's difficult to stop something when all the nurses and orderlies are looking on." Ibid.

121. Ibid.

122. Ibid.

123. *In Re Quinlan*, 664–65, 671 (emphasis added).

124. Cf. Robert M. Veatch, *Death, Dying and the Biological Revolution* (New Haven: Yale University Press, 1976, 1989), 122–23.

125. Quinlan, Quinlan, and Battelle, *Karen Ann: The Quinlans Tell Their Story*, 280. Additionally, ethics committees were not taken seriously by the medical community until President Reagan's intervention in the "Baby Jane Doe case."

126. Ibid.

127. Rothman, *Strangers at the Bedside*, 222. See also for example, Robert Veatch, *Death, Dying and the Biological Revolution*, rev. ed. (New Haven: Yale University Press, 1989), 120; Cathie Lyons, "The Quinlan Decision," *Christianity and Crisis*, May 10, 1976, 103–4.

128. Letter to the Editor, *Los Angeles Times*, November 18, 1975, sec. 2, 6.

129. Kenneth Vaux, "Beyond This Place: Moral-Spiritual Reflection on the Quinlan Case," *Christian Century*, January 21, 1976, 46.

130. Quinlan, Quinlan, and Battelle, *Karen Ann: The Quinlans Tell Their Story*, 252–53.

131. See, for example, "The Right to Live—or Die," 40–45; "The Quinlan Decision," 107.

132. "The Right to Live—or Die," 40; "The Quinlan Decision," 107.

133. See, for example, the comments of Andre Hellegers of the Kennedy Institute and Robert Veatch of ISELS quoted in Editorial, *Science News*, October 4, 1975, 213. But see discussion of George Annas (below) as a notable exception.

134. See, for example, Powledge and Steinfels, "Following the News on Karen Quinlan—a Media Watch on a Landmark Case," 5, and Roy Branson, Kenneth Casebeer, Melvine D. Levine, Thomas C. Oden, Paul Ramsey, and Alexander Morgan Capron, "The Quinlan Decision: Five Commentaries," *Hastings Center Report* 6, no. 1 (February 1976): 8–19; Andre Hellegers, "Quinlan: No Sweeping Precedent," *Ob-Gyn News* 11, no. 15 (August 1, 1976): 16; and Robert Veatch, "The Quinlan Case: A Tragedy of Faulty Analysis," *Bioethics Northwest* 1, no. 1 (winter 1976): 2+.

135. Thomas C. Oden, "Judicial Restraint in the Quinlan Decision," *Christian Century*, November 26, 1975, 1068–69.

136. Kenneth Briggs, "The Fragile Issue of 'Death Control'," *New York Times*, October 13, 1975, 33.

137. George Annas, "In Re Quinlan: Legal Comfort for Doctors," *Hastings Center Report* 6, no. 3 (1976): 29–31. See also Richard A. McCormick, S.J., "The Karen Ann Quinlan Case," *JAMA* 234, no. 10 (December 8, 1975): 1057, which discusses the liability concerns which contributed to causing physicians' refusal to terminate treatment.

138. Willard Gaylin, Cathie Lyons, Harmon L. Smith, Roger L. Shinn, and Robert M. Veatch, "Who Should Decide? The Case of Karen Quinlan," *Christianity and Crisis*, January 19, 1976, 322–30.

139. See Rothman, *Strangers at the Bedside*, 244–46. For an analysis distinct from Rothman's, see Renee Fox, "The Sociology of Bioethics," in *The Sociology of Medicine: A Participant Observer's View*, ed. Alex Inkeles (Englewood Cliffs, N.J.: Prentice-Hall, 1989), especially 230–31. Also see "Too Much Law?" *Newsweek*, January 10, 1977, 42, and Stuart A. Scheingold, *The Politics of Rights* (New Haven: Yale University Press, 1974).

140. Shannon, "A Triumph of Technology," 589.

141. Gaylin et al., "Who Should Decide? The Case of Karen Quinlan," 323–24.

142. Vaux, "Beyond This Place," 43–44.

143. See, for example, Paul Ramsey, *Ethics at the Edges of Life* (New Haven: Yale University Press, 1978); Thomas L. Beauchamp and James F. Childress, *Principles of Biomedical Ethics* (New York: Oxford University Press, 1979); Veatch, *Death, Dying and the Biological Revolution*.

144. Quinlan, Quinlan, and Battelle, *Karen Ann: The Quinlans Tell Their Story*, 284, 285.

145. Cf. Rothman, *Strangers at the Bedside*, 223.

Epilogue

1. Alvin Toffler, *Future Shock* (New York: Bantam Books, 1971), 205, 438, 2.

2. The development of the Office of Technology Assessment during this time is a related phenomenon.

3. Portions of this epilogue are reprinted with permission from my article, "What *Quinlan* Can Tell Kevorkian about the Right to Die," *Humanist* 57, no. 2 (1997): 10–14.

4. Robert Truog, "Is It Time to Abandon Brain Death?" *Hastings Center Report* 27, no. 1 (1997): 29–37. Truog's analysis and recommendations, themselves part of the ongoing debate on how to define death, are not undisputed. See, Letters to the Editor, *Hastings Center Report* 27, no. 1 (1997), and James L. Bernat, "A Defense of the Whole-Brain Concept of Death," *Hastings Center Report* 28, no. 2 (1998): 14–23.

5. See Robert M. Veatch, "The Whole-Brain-Oriented Concept of Death: An Outmoded Philosophical Formulation," *Journal of Thanatology* 3 (1975): 13–30; Veatch, "The Impending Collapse of the Whole-Brain Definition of Death," *Hastings Center Report* 23, no. 4 (1993): 18–24.

6. Truog, "Is It Time to Abandon Brain Death?" 33.

7. James L. Bernat, "A Defense of the Whole-Brain Concept of Death," *Hastings Center Report* 28, no. 2 (1998): 14–23, 22. In addition to Bernat's critique, see Letters to Editor, in *Hastings Center Report* 27, no. 5 (1997), for examples of practitioners who disagree with Truog and for Truog's response to critics.

8. Truog, "Is it Time to Abandon Brain Death?" 29–37.

9. Ibid., 33. See also, Robert M. Arnold and Stuart J. Younger, "The Dead Donor Rule: Should We Stretch It, Bend It, or Abandon It?" *Kennedy Institute of Ethics Journal* 3 (1993): 263–78.

10. See Guy McKahn, "The Modern Meaning of Death," *Washington Post*, October 25, 1998, C1, for an example of the debate's relatively rare exposure in a more popular venue. See also Jim Holt, "Sunny Side Up," *New Republic*, February 21, 1994, 23, for an example of popular understanding of how definitions of death are considered.

11. *Cruzan et ux. vs. Director, Missouri Department of Health*, 497 U.S. 261, 110 S. Ct. 2841 (1990).

12. *Washington vs. Glucksberg*, 521 U.S. 702 (1997), 117 S. Ct. 2258 (1997); *Vacco vs. Quill*, 521 U.S. 793 (1997), 117 S. Ct. 2293 (1997).

13. Cf. www.religioustolerance.org/euthanas.htm. For further analysis of these cases, see Alexander Morgan Capron, "Death and the Court," *Hastings Center Report* 27, no. 5 (1997): 25–29.

14. *Washington vs. Glucksberg*, 521 U.S. 702 (1977) at p. 735.

15. Or. Rev. Stat. ss 127.800 et seq. (1997) as amended by 1999 Ore. SB 491.

16. See also "Poll Shows Strong Support for Assisted Suicide," *San Francisco Chronicle*, July 30, 1998, A6.

17. See, for example, Harry van Bommel, "Dying Well," *Next City* 3, no. 3 (Spring 1998): 36.

18. Derek Humphry and Mary Clement, *Freedom to Die: People, Politics and the Right-to-Die Movement* (New York: St. Martin's Press, 1998), 333.

19. John Hardwig, "Is There a Duty to Die?" *Hastings Center Report* 27, no. 2 (1997): 34–42, at 34.

20. See, Wesley J. Smith, "Assisted Suicide Isn't 'Death With Dignity,'" *San Francisco Chronicle*, November 12, 1998, A27, for a discussion of problems associated with supporting physician-assisted suicide in a society where cost-conscious health maintenance organizations dominate medical practice.

Index